Evolutionary Moral Realism

Against standard approaches to evolution and ethics, this book develops the idea that moral values may find their origin in regularly recurring features in the cooperative environments of species of organisms that are social and intelligent.

Across a wide range of species that are social and intelligent, possibilities arise for helping others, responding empathetically to the needs of others, and playing fairly. The book identifies these underlying environmental regularities as biological natural kinds and as natural moral values. As natural kinds, moral values help to provide more complete explanations for the selection of traits that arise in response to them. For example, helping in an aquatic environment is quite different than helping in an arboreal environment, and so we can expect the selection of traits for helping to reflect these underlying environmental differences. With the human ability to name, talk, and reason about important features of our environment, moral values become part of moral discourse and argument, helping to produce coherent systems of moral thought.

Combining a naturalistic approach to morality with an equal emphasis on moral argument and truth, this book will be of interest to philosophers and historians of biology, theoretical biologists, comparative psychologists, and moral philosophers.

John Collier (1950–2018) was Professor Emeritus at the University of KwaZulu-Natal, South Africa.

Michael Stingl is Associate Professor of Philosophy at the University of Lethbridge, Canada.

History and Philosophy of Biology

Series Editor: Rasmus Grønfeldt Winther

Associate Professor of Philosophy at the University of California, Santa Cruz (UCSC)

This series explores significant developments in the life sciences from historical and philosophical perspectives. Historical episodes include Aristotelian biology, Greek and Islamic biology and medicine, Renaissance biology, natural history, Darwinian evolution, Nineteenth-century physiology and cell theory, Twentieth-century genetics, ecology, and systematics, and the biological theories and practices of non-Western perspectives. Philosophical topics include individuality, reductionism and holism, fitness, levels of selection, mechanism and teleology, and the nature-nurture debates, as well as explanation, confirmation, inference, experiment, scientific practice, and models and theories vis-à-vis the biological sciences.

Authors are also invited to inquire into the "and" of this series. How has, does, and will the history of biology impact philosophical understandings of life? How can philosophy help us analyze the historical contingency of, and structural constraints on, scientific knowledge about biological processes and systems? In probing the interweaving of history and philosophy of biology, scholarly investigation could usefully turn to values, power, and potential future uses and abuses of biological knowledge.

The scientific scope of the series includes evolutionary theory, environmental sciences, genomics, molecular biology, systems biology, biotechnology, biomedicine, race and ethnicity, and sex and gender. These areas of the biological sciences are not silos, and tracking their impact on other sciences such as psychology, economics, and sociology, and the behavioral and human sciences more generally, is also within the purview of this series.

Ecological Investigations
A Phenomenology of Habitats
Adam C. Konopka

Evolutionary Moral Realism
John Collier and Michael Stingl

For more information about this series, please visit: www.routledge.com/History-and-Philosophy-of-Biology/book-series/HAPB

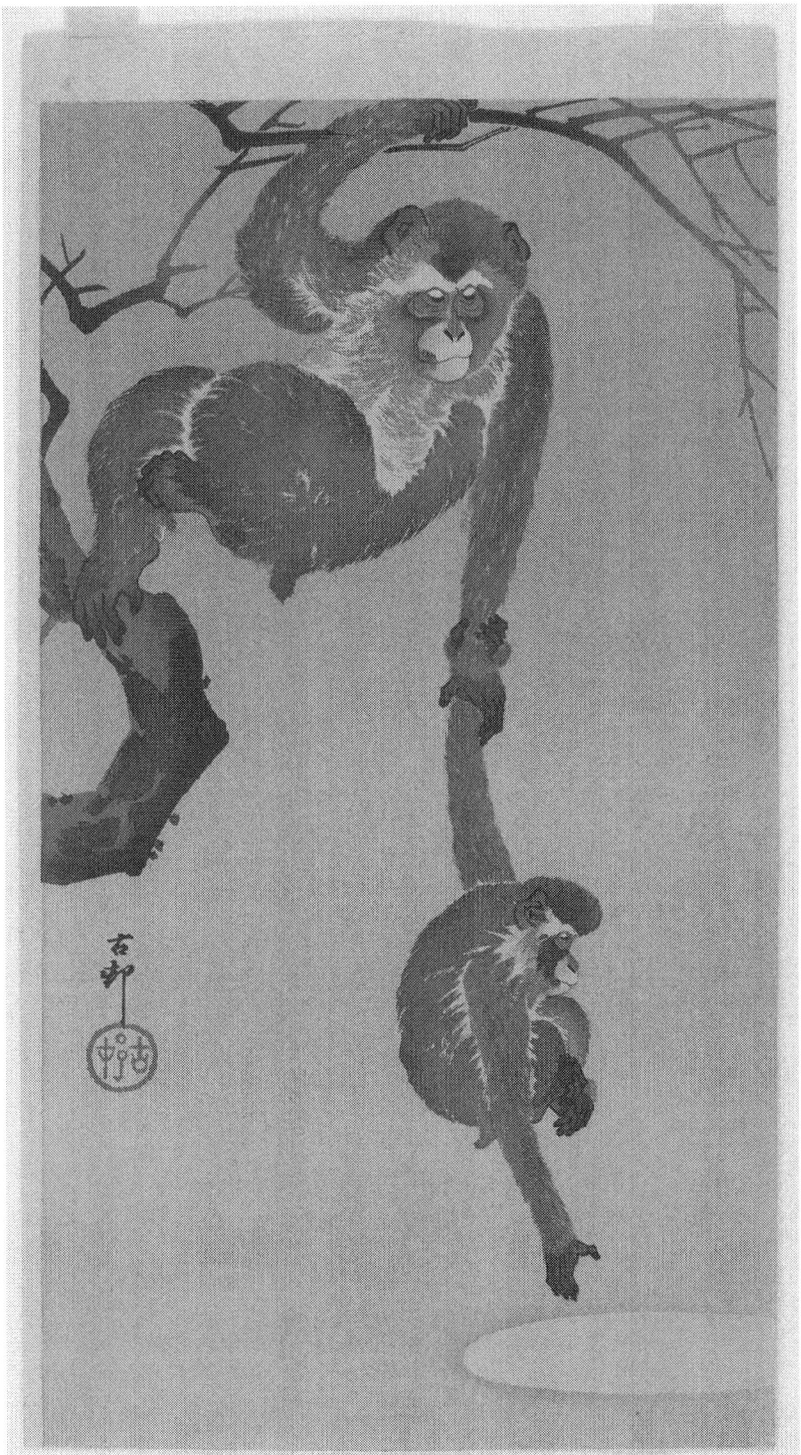

Koson, Mother and baby reaching for the reflection of the moon. Permission granted by the Arthur M. Sackler Gallery, Smithsonian Institution, Washington, DC, Robert O. Muller Collection.

Evolutionary Moral Realism

John Collier and Michael Stingl

Routledge
Taylor & Francis Group
LONDON AND NEW YORK

First published 2020
by Routledge
2 Park Square, Milton Park, Abingdon, Oxon OX14 4RN

and by Routledge
52 Vanderbilt Avenue, New York, NY 10017

Routledge is an imprint of the Taylor & Francis Group, an informa business

British Library Cataloguing-in-Publication Data
A catalogue record for this book is available from the British Library

Library of Congress Cataloging-in-Publication Data
Names: Collier, John (Philosophy), author. | Stingl, Michael, 1955- author.
Title: Evolutionary moral realism / John Collier and Michael Stingl.
Description: Abingdon, Oxon ; New York, NY : Routledge, 2020. | Series: History and philosophy of biology | Includes bibliographical references and index.
Identifiers: LCCN 2019039353 (print) | LCCN 2019039354 (ebook) | ISBN 9780367281304 (hbk) | ISBN 9780429299803 (ebk)
Subjects: LCSH: Ethics. | Evolution (Biology) | Evolutionary developmental biology.
Classification: LCC BJ58 .C65 2020 (print) | LCC BJ58 (ebook) | DDC 171/.7—dc23
LC record available at https://lccn.loc.gov/2019039353
LC ebook record available at https://lccn.loc.gov/2019039354

ISBN: 978-0-367-28130-4 (hbk)
ISBN: 978-0-429-29980-3 (ebk)

Typeset in Times New Roman
by Apex CoVantage LLC

In memory of John

Contents

	About the authors	x
	Acknowledgements	xi
	Preface	xiii
1	Evolutionary moral realism	1
2	The moon in the water	17
3	Moral trajectories	39
4	Moral sense theories	60
5	Reason, rational contracts, and selfish genes	83
6	Natural moral values and moral progress	109
7	Partial and impartial moral reasons	129
8	Moving from is to ought	144
	Conclusion	167
	Index	171

About the authors

John Collier (1950–2018) was Professor Emeritus at the University of KwaZulu-Natal. He received his PhD from The University of Western Ontario and taught or held research positions at Rice, Calgary, Newcastle, Melbourne, and the Konrad Lorenz Institute. His research interests were in the philosophy of science, biological and evolutionary theory, information theory, and complexly organized systems. He published widely in journals such as *Biological Theory*, *Biosemiotics*, *Cognition, Communication and Co-operation*, *Biosystems*, *Theoria*, *The Australasian Journal of Philosophy*, *Revue Internationale de Philosophie*, and *Studies in the History and Philosophy of Science.*

Michael Stingl received his PhD from Toronto and has taught at Rice, Calgary, and the University of Lethbridge, where he is currently an associate professor of philosophy. He has sat on several provincial health ethics boards and was the coordinator of the editorial board of the Canadian Journal of Philosophy. His research interests are in the areas of evolutionary ethics and biomedical ethics. He has published articles in *Biological Theory, Biology and Philosophy*, *The Canadian Journal of Philosophy*, *Ethical Theory and Moral Practice*, and *The Cambridge Quarterly of Healthcare Ethics*. He is the editor of *The Price of Compassion: Assisted Suicide and Euthanasia* (2010).

Acknowledgements

This book has been a long time in the writing. The project began with a paper we published together in 1993 on evolutionary naturalism and the objectivity of morality. Since then, we published several other papers on the view we came to call evolutionary moral realism (EMR), with the longer range intention of producing a book manuscript. A draft of the book was completed in 2018, shortly before John died after a period of steadily deteriorating health.

I had been doing most of the writing and continued on with the editing tasks that have now produced this book. From 1993 to 2018, we exchanged a steady stream of e-mails and met regularly to work on the book. By the time we had completed the draft manuscript, we had agreed upon two things about it. First, as we worked out the idea of EMR, what each of us added or subtracted seemed to be only what the theory itself required. Second, in terms of what each of us was adding and subtracting, neither of us could have written this book without the other. If one of us had what seemed like a key idea, the other usually saw why or why not. While there is some of each of us in the book, there is a lot of both of us.

John was avid in his conference attendance and in his correspondence with others. He no doubt would have had his own list of people to acknowledge, a list I can no longer ask him for. So I will list those whom I know about from both our experiences in working with the book material.

Three people read the manuscript as a whole at various stages in its development. Bill Rottschaefer read an earlier version and responded with some very helpful comments. Lucas McGranahan worked through a later version as a developmental editor. His comments and suggestions did much to help clarify the storylines of the individual chapters and the book as a whole. Thanks are due as well to Rasmus Gronfeldt Winther, series editor, for prodding me to improve the manuscript in some very useful directions.

Others read bits of the book in one form or another and had useful comments for us: Bryson Brown, Richmond Campbell, Julia Clare, Abe Gibson, Trudy Govier, Lynn Kennedy, Karl Laderoute, Bert Musschenga, Kent Peacock, Sergio Pellis, Ronnie de Sousa, and Ute Wieden-Kothe. We also received useful feedback from a number of anonymous referees over the years, from audience members when one or the other of us presented material from the book at conferences, and from students in our classes at the University of Lethbridge and the University of Kwazulu-Natal.

We would like to thank the Arthur M. Sackler Gallery, Smithsonian Institution, Washington, DC, Robert O. Muller Collection, for permission to use the Koson block print of the two monkeys reaching for the moon as the frontispiece for the book.

My departmental colleagues Bryson Brown and Kent Peacock have been steadfast supporters of this project from early on, and both John and I profited much from ongoing conversations with each of them. John profited in similar ways from working on biological complexity with Mishtu Banerjee, Dan Brooks, Werner Callebaut and other colleagues at the Konrad Lorenz Institute, Cliff Hooker, Jack Maze, Konrad Talmont-Kaminski, and Ed Wiley.

Ken Nummela was a lifelong and caring friend of John's. Penny Fabian was a more recent and geographically closer friend, one who provided significant emotional support to John near the end of his life. I am sure he would have wanted to thank both of them for their friendship and their support.

My partner, Jacqueline Preyde, in addition to listening to innumerable arcane discussions of natural and moral kinds, proofread and helped to copy-edit the final version of the manuscript. At crucial junctures she pushed me to keep working on it. Her love and support have been invaluable to me.

Michael Stingl

Preface

In the early 1990s, when we published our article "Evolutionary Naturalism and the Objectivity of Morality" (Collier and Stingl, 1993), we found ourselves growing increasingly dissatisfied with approaches to evolution and ethics that treated humans like mushrooms freshly sprung. We were similarly sceptical of the idea that morality was an artefact of the human mind and that it was, as such, an exclusively human phenomenon. Our feeling was that something biologically interesting and important was keeping humans together in closely cooperative groups before they got smart, articulate, and in the business of constructing increasingly complex social norms and structures that defined their expectations of and obligations to their fellow group members and other related individuals, such as trading partners or animistic spirits. We also thought that whatever this biologically interesting and important something was, it predated our particular species line. It was likely to have started earlier, and perhaps much earlier.

The main problem defining more standard approaches to squaring ethics with evolution seemed to us to get matters the wrong way around. The question of how evolution might have gotten self-oriented individuals into cooperative groups that limited individual goods in the interest of more common goods seemed to assume that such closely cooperative individuals could have somehow arisen on their own rather than as interlocking parts of more complex and dynamic evolutionary processes that would have connected them and their individual goods tightly together from the very beginning. Humans, we thought, were never likely to have arisen as mushrooms freshly sprung.

In Collier and Stingl (1993), we argued for the theoretical possibility of an alternative view of evolution and ethics, a view that took moral values to be embedded elements in processes of evolutionary development that led to intelligent and social species. In this book, we argue not just for the possibility of this view but for its plausibility. Currently, there is not enough evidence available for us to argue more conclusively that this view is more likely than its competitors to be on the track of the empirical truth about evolution and ethics, but we do think there is now enough evidence available to suggest that it would be a very good idea to start looking for more.

Since 1993, evidence for thinking that a more dynamical approach to evolution and ethics is not just possible but also plausible has been accumulating on several fronts. At the biochemical level, what seems to be at the centre of evolutionary processes are not master-molecule genes but dynamic systems of interacting elements where the well-functioning of each part of the system is closely connected to the well-functioning of other parts. At the environmental level, ideas about the well-functioning of ecosystems have been rapidly moving in the same direction, to more dynamical and complex models linking the individual goods of members of ecosystems with each other and with more systemic goods. Finally, in comparative psychology, other animals seem to be responding to features of their environments that we humans would label in our own environment as morally interesting and important things, such as responding positively to opportunities to help others in trouble, empathetically caring for others when they are distressed, and interacting fairly with others in exchanging various goods or favours.

That such pro-social traits have arisen across such a wide variety of species suggests to us that there might be underlying environmental regularities driving the selection of these traits. This suggests that if the traits are biologically interesting, the underlying regularities structuring the selection processes behind the evolutionary development of these traits are even more interesting, as well as more important. In thinking about biological systems at the level of biochemical interactions and at the level of social interactions between individual organisms, we have increasingly come to think that we are looking at developing systems containing individual parts that could never have arisen on their own, like mushrooms freshly sprung. Humans are not like mushrooms freshly sprung, nor are genes. Not even mushrooms are like mushrooms considered in this sort of way.

Pushing against the view we are developing in this book are several strong currents of thought. The most recent, but also the most tenuous, is the idea of selfish genes, linked to the ideas of kin selection and reciprocal altruism. This way of thinking about evolution and ethics is rooted more deeply in pervasive forms of twentieth-century positivistic and scientific thinking that would locate the natural source of values in attitudes and emotions rather than in cognitive responses to independently existing features of the biological world. Such thinking is rooted even more deeply in the unfolding legacy of the early modern revolution in Western thought that produced rational self-interest, liberal individualism, and the idea that everything in the natural world must have a fully mechanistic explanation. With the rise of modern science, values, along with spirits, have increasingly seemed to have no natural place in the natural world, at least not as independently existing and causally efficacious parts of it.

Pushing back against these currents of thought, the purpose of this book is twofold: first, to argue, empirically, that it is plausible to suppose that there may be certain systemic regularities in the environments of social and intelligent species that form a natural kind or closely linked group of natural kinds; and second, to argue, philosophically, that if these kinds exist, they are likely to be moral kinds, what we have come to call "natural moral values." While we are not in a position to offer conclusive versions of either argument, we do think that there are good reasons for supposing that such an approach to evolution and ethics might be much more plausible than current thinking about evolutionary ethics might otherwise lead us to believe.

This result is not inconsequential. If our arguments are on the right track, they lead in the direction of an evolutionary ethics that makes moral values deep and pervasive aspects of the environments of any species that is social and intelligent. In helping to drive the selection of traits that could then be explained as arising in response to them, these values would be causally real parts of the natural world. From the point of view of biological theory, they would similarly provide a way to integrate at a more general level the explanations of a wide variety of traits across a wide variety of species.

More philosophically, if our arguments are on the right track, morality is not a negligible aspect of the biological world, arising late in the evolutionary day with the appearance of humans. It is a deeply significant aspect of the evolution of species that are social and intelligent, from near their very beginnings. Morality is not simply an artefact of human capacity for thought, language, and argument, but, instead, moral arguments are aimed, at their foundation, at moral values that exist independently of our human ability to comprehend them.

A guiding idea of our arguments is that pro-social emotions, in humans and other animals, evolved in response to the same general kinds of features in the social environments of the species involved. This happened before humans developed the capacity to talk, think, and argue about these features as they came to be realized in their own particular environment. Once the later capacities appeared, the earlier features of the human environment, the features we identify in this book as natural moral values, became some of the most important things that our biological ancestors found themselves talking, thinking, and arguing about.

This certainly does not mean that there is one true system of ethical thought, one that all humans will ultimately agree upon; but it does mean that human systems of ethical thought are likely to be tracking, more or less closely, the same underlying moral values that ground these systems of thought at their shared biological base. On this biological approach to morality, human morality becomes a special case of a broader biological phenomenon. As part of species becoming social and intelligent, certain features of their environments become natural attractors, or positive moral values, things like helping others in trouble, empathetically caring about the distress of others, and treating others fairly. Correspondingly, certain features of such environments become naturally problematic, like cheating. To some degree cheating is naturally attractive, but to some degree it is also naturally repellent. As humans, we talk and argue about moral values, both positive and negative. If the arguments of this book are on the right track, an important part of what we are doing with such talk and argument is trying to figure out how best to respond to natural moral values as they appear to us in our own culturally evolving set of social circumstances. While some of our moral conclusions may track these values more closely, others may not. Not all moral values need be supposed to be natural moral values, but at the core of all moral systems, we should expect to find natural moral values.

If, that is, the arguments of this book are on the right track. Let us see if they are.

Bibliography

Collier, John, and Michael Stingl. 1993. "Evolutionary Naturalism and the Objectivity of Morality." *Biology and Philosophy* 8:47–60.

1 Evolutionary moral realism

Evolutionary biology and the nature of morality

"If ye break faith with us who die / We shall not sleep, though poppies grow / In Flanders fields." These lines from a well-known poem by Robert McCrae, a Canadian physician, poet, and soldier in the First World War, suggest that breaking faith with those who have sacrificed themselves so that we may survive and prosper is a bad thing to do. It is, the poem suggests, morally wrong. Maintaining bonds of social trust in such circumstances – and remaining loyal to the cause for which the sacrifice was made – is morally good. Many of us would share this intuition.

But where do beliefs about moral values come from, and what makes them correct or incorrect? We contend that morality may be a general biological phenomenon grounded in the evolution of intelligent, social species. Morally bad actions generally lead to unhappiness and misery, and morally good ones generally have opposite effects on human well-being, both individually and socially. These effects are generally linked to further biological consequences, such as survival and reproduction. Nevertheless, nothing is morally good simply because it is the product of evolution. This book develops a naturalistic approach to morality that we call evolutionary moral realism (EMR) in Collier and Stingl (2013). We begin with the intuitive idea that certain kinds of things existing in an organism's environment may be naturally good for that organism before the organism develops any sort of a capacity to detect or respond to them. We focus on one group of naturally good things that we call natural moral goods, and we pay particular attention to one kind of organism, humans, for which normatively robust moral ought-claims seem appropriate. But we also consider other kinds of organisms that we think are capable of detecting natural moral goods that arise in their environments, sometimes in ways that enable these organisms to respond in morally appropriate ways. Moral goods may guide the behaviour of such organisms, and so these goods may have a normative function, even if it is incorrect to say that the organisms are following explicit norms that tell them how they ought to behave.

Unlike most other evolutionary approaches to ethics, EMR begins with an empirical hypothesis that invokes a particular form of moral realism. Loosely speaking, moral values are morally good or morally bad kinds of things that regularly arise in the environments of species that are social and intelligent. Morality is not ultimately based in behaviour patterns that have survival value or in psychological capacities that might lead humans to believe that certain things are morally good or morally bad. EMR's theoretical commitment to moral realism is not intended as a metaphysical postulation but as the guiding hypothesis of an empirical research program that we argue is both plausible and interesting. We should be clear from the outset that all we are trying to establish in this book is the initial plausibility and theoretical interest of this hypothesis. More conclusive arguments in its defence will require more empirical development. But this is as it should be: we are advancing EMR as an empirical theory of morality, and at this early point in its development, the very beginning of such a theory.

A key question the book seeks to answer is how we might get from EMR's hypothesis about natural moral goods to the ought-claims of human moral codes. Drawing together arguments from earlier chapters, our final chapter focuses on the empirical relationship between what we are calling natural moral facts and normative moral values, arguing that EMR makes no logically fallacious move from "is" to "ought." The nature of the relationship from "is" to "ought" is central to EMR's success as an empirical theory of morality, because we can only be sure that we are offering an account of what we humans call "morality" if EMR is able to offer plausibly linked accounts of moral justification and moral truth. To get to the endpoint of moral justification and moral norms in the way that we do, we start empirically with the idea that morally good things, as well as morally bad things, may initially arise in the biological world as natural kinds of things in the environments of certain sorts of species.

Consider a biological kind like a lion. This kind of thing appears as a regularly recurring structural feature of the environments of certain other kinds of creatures, some of which may benefit greatly from being able to detect this particular feature of their environment and to respond to it appropriately, that is, as a predatory kind of thing.

According to EMR, we should suppose something similar for moral goods: they arise in the material world as natural biological kinds, something like regularly recurring predatory patterns like lions. We do not, however, think that moral kinds are species kinds, a point we will clarify in Chapters 2–4. Our point here is that in their simplest forms, natural moral goods are regularly recurring structural features of particular environments, features that might come to matter greatly to the organisms in those environments. To see what we mean here, consider three quick examples.

Build a rat trap that rats can be trained to open. Put an untrained rat in the trap and then in a cage place the trapped rat, a trained rat, and some food. The trained rat will often free the trapped rat before both rats then eat the food (Bartal, Decety, and Mason 2011).

If you are drawing a picture with a marker and drop it where you cannot reach it, both small children and chimps will hand the marker back to you. They will not do this if you have thrown the marker down (Warneken et al. 2007).

Small children are watching a puppet that is apparently trying to get a brightly coloured rattle out of a box. A second puppet sits on the lid of the box so that it cannot be opened. A third puppet helps to open the lid of the box. If you offer the latter two puppets to the children, they are most interested in the puppet that helped (Hamlin and Wynn 2011).

Although it may be realized in different ways, the possibility of helping someone in need would seem to be a regularly recurring structural pattern in the environments of rats, chimps, and humans. This pattern is interesting and important to these creatures. Just as it is good to be able to detect and respond to predatory patterns in their environments, it is good for organisms like rats, chimps, and humans, and for their genes, for them to be able to detect and respond appropriately to natural moral goods. Consider two other experiments.

If you reward capuchin monkeys with highly desirable grapes when other capuchins have been receiving less desirable cucumbers for the same task of returning tokens to you, the monkeys getting the cucumbers will stop cooperating. They will refuse the cucumber, throwing it on the ground or back at you (Brosnan and de Waal 2003).

In play fighting, juvenile rats follow a fifty-fifty rule in knocking each other down and going after each other's throats (Pellis, Pellis, and Reinhart 2010, 405–406). The rats stop playing with playmates who do not allow them an equal chance to fight back, and as adults, rats left out of such play tend to behave too aggressively in unthreatening social circumstances and not aggressively enough in threatening circumstances.

The possibility of playing fairly, like helping someone in need, is a naturally recurring pattern in the environments of certain sorts of creatures. These kinds of good things arise in similar environments and are connected to other kinds of good things, like trusting and caring relationships, in mutually supporting kinds of ways. EMR calls such things natural moral goods. It may be that the natural moral goods group together as a single, more general kind of thing, or alternatively, that they form a matrix of closely related kinds of things. Either way, they are what EMR calls natural moral goods. Patterns of things that are morally bad, like cheating, may form a similar class of natural moral values, in this case negative moral values.

EMR thus treats human morality as a particular instance of a more general biological phenomenon. Things that are morally good for us are also good for other kinds of organisms, and some of these organisms are able to respond to these goods in ways that are appropriate to the kind of good thing that they are. Humans evolved to be able to talk and argue about these good kinds of things, and such talk and argument has led, culturally and historically, to carefully articulated sets of moral norms.

This evolutionary approach to morality immediately raises three important questions. First, what makes morally good things morally good, on this naturalistic account? Second, why does moral goodness, on this approach to morality, not simply reduce to what is good for our genes? And third, how exactly are natural moral goods supposed to be related to moral oughts? In particular, does EMR commit the logical fallacy of inferring normative claims from factual claims? In the chapters to follow we offer extended answers to these questions on behalf of

EMR. Given their importance to the general argument of the book, we begin here with a sketch of how we intend to answer them.

What makes natural moral goods morally good?

Nothing. EMR's answer to this question may initially seem disappointing. According to EMR, natural moral goods are nothing more than the naturally occurring good kinds of things that they are. As in the famous epigraph of G.E. Moore's (1903) *Principia Ethica*, everything is what it is and not something else. If EMR is right, it turns out that at its beginning, morality is a general biological phenomenon. Some things are morally good and others are not. Helping another and playing fairly are what that they are, and there is no more foundational a source of moral goodness to tell us why this is so. Moral goods may have some sort of biological function, such as regulating conflicts between individual and group goods, but even if this is part of what makes them the natural kinds of things that they are, it is not what makes them moral kinds.

More metaphysically inclined philosophers like Moore might still ask: supposing that the kinds of things in question exist as natural kinds, what is it about them that makes them *moral* kinds? To such philosophers, it might look as if EMR is simply begging the question in favour of its evolutionary answer to the question, What is morality? In some ways, we may be begging this question, but not, we think, in any way that is fatal to the success of EMR.

We aim to make plausible the idea that morality may ultimately turn out to be a particular natural kind or a cluster of closely related natural kinds. So we do face an immediate challenge in the form of Moore's open question argument: how do we know that this natural kind, whatever it turns out to be, is a *moral* kind? If this is indeed an open question, EMR has yet to tell us what morality really is. But our response to this challenge is just as immediate: the open question argument only works if we assume "morality" is not originally a directly referential natural kind term. We can ask, philosophically, if water really is H_2O; but if we have an empirical theory that tells us that what we call "water" does indeed turn out to refer to H_2O, the philosophical openness of the question about what water "really is" loses much of its point. "Water" might still have important connotations not directly tied to its reference, but given the kind of thing it refers to, any cultural connotations humans might want to give it will need to take its referential meaning into account.

In other words, the fundamental meaning of morality might already be fixed by its objective referent in the world. If we defend EMR in this way, we may still appear to be open to the philosophical objection that we are simply assuming "morality" is in its most basic use a natural kind term. There are plenty of philosophical arguments that morality is something else, or even nothing at all, other than a cognitive and emotional illusion that evolution has built us humans to fall prey to.

These alternative approaches to morality may of course raise open questions of their own. Nevertheless, what are we to say about the problem for EMR? Our response to EMR's version of the open question argument is that we are not simply assuming that key moral terms are directly referential but actively developing an empirical theory of morality that treats them as if they were. We think the theory has enough initial plausibility to be taken seriously and that it may be empirically testable. In this book, we are not yet in a position to state the theory fully or to test it. But if we can establish that EMR has some degree of initial plausibility, philosophical arguments that EMR is unnecessary or impossible become correspondingly less interesting as theories of morality. Existing approaches to morality are mostly if not entirely speculative, they do not agree with one another, and they are not testable in their current forms. Our development of EMR in this book does not provide specific arguments that might refute any of them, but it does promise to stay within the bounds of empirical testability.

In developing our argument for EMR, we will not be reviewing competing approaches to morality, one at a time, trying to show where and how they might be mistaken. They might not be. EMR may instead be mistaken, but if it is, the mistake will be an empirical one. On the other hand, if we can show that EMR has some level of empirical plausibility, taking on other speculative arguments about what morality might be does not seem like a particularly pressing task at this early point in EMR's development.

A central claim of EMR is that rats, monkeys, chimps, and small children can detect and respond to moral goods when they are present in their environment. They cannot of course name these kinds of things or talk or argue about them. They cannot detect ways in which the different goods might be connected to one another in mutually reinforcing ways, and they cannot generalize from observations of the goods across different contexts, either as the goods might separately arise or as they might arise in conjunction with each other.

This might seem to be an immediate and significant empirical problem for EMR. On most other evolutionary approaches to ethics, morality begins with the human capacity to name, talk, and argue about aspects of our social interactions with one another. "This is (morally) good" and "do (or promote) this" mark out aspects of our social situation to which enough of us are willing to extend joint approbation. In the beginning was the Word, and the Word was Good. What makes things morally good is our capacity and willingness to recognize them as such, not the fact that they are good in themselves. The target, then, for most other approaches to evolutionary ethics, their explanandum, is to explain why we have such tendencies to group together and promote these nominal moral values. This human tendency may be related to more rudimentary psychological capacities in other species, but there is no such thing as morality in the world until humans evolve to create it.

EMR turns this idea on its head. In the beginning were the naturally arising moral goods. The Word came much later, with the human development of language and argument; and if the Word was indeed good, this was at least in part because it allowed humans to talk and argue about naturally occurring moral goods. In the beginning, these goods made it advantageous, in certain environments, to be able to detect them in ways that were directly connected to behaviour.

Once the goods appeared in the environment of a creature, capacities to detect and respond appropriately to them became possible, given the requisite biological variation arising within the creature's existing developmental constraints. If the environment favoured such capacities, they could be selected for. Initial capacities might lead to more discriminating capacities and perhaps to new forms of moral goodness. In the right sorts of environments, more discriminating capacities might disclose aspects of moral goodness invisible to less discriminating capacities, and more discriminating capacities might create new social possibilities that might bring with them new forms of moral goodness. In such a virtuous feedback loop, moral goods might affect trait selection, and in affecting trait selection, they may further affect group selection including species selection.

EMR takes moral goods to be causally significant structural features of the biological world that help explain the various kinds of capacities that arise in response to them. This includes how and why similar capacities arise in species that are not closely related to one another (convergent evolution) or differ across related species. In treating human morality as a specific instance of a more general biological phenomenon, EMR is offering a hypothesis that unifies and simplifies at a more general level related explanations of what appear to be similar behaviour patterns across a wide variety of species.

Why not suppose, perhaps even more simply, that the capacities arose all by themselves, without reference to anything objectively good? In a well-known thought experiment, Gilbert Harman (1977, 4) asks us to imagine that we have just come across a group of hooligans engaged in setting a cat on fire. In this scenario, Harman asks us where it is simpler to locate the apparent moral badness of the situation: out in the material world with the match and the flames and the screaming cat, or in our own inner revulsion at the hooligan's delight in the screaming cat engulfed by flames? The apparent badness, says Harman, is most simply accounted for in terms of our emotional response. We simply do not like this kind of thing. And if we look back to human evolution, we can see how it would have been beneficial to social and intelligent creatures like us to have exactly this sort of response, or at least enough of us enough of the time, since evolution also seems to make hooligans possible.

EMR aims to provide a more general explanatory account of morality and moral responses than an empirical account that focuses on responses alone. The key idea is that species-specific moral capacities are themselves patterned by a more general underlying moral and biological reality. Moral values are regularly recurring structural features of the environments in which particular moral capacities develop. While what exactly it is to help another may vary in detail across the environments of primates, corvids, and porpoises, helping itself may be one and the same general kind of thing.

In an extension of Harman's argument, Sharon Street (2006) argues that while we should suppose that moral emotions arose because they were linked to behaviours that had selective advantages, we should also suppose that they would have been linked to real features of the environment in which they evolved. Something has to trigger the emotions in environments where the moral behaviours pay off.

Occam's razor would seem to caution us not to suppose that these real features of the environment would be mysterious entities like moral values – much better to suppose that they would just be real things in the human environment like the screams of cats.

According to EMR, Street's argument is telling us to look for the wrong kinds of things, or perhaps more fairly, not to look for the right kinds of things in the right kind of way. What EMR tells us to look for are general structural patterns in the environments of certain kinds of creatures; these environmental patterns then pattern adaptations that are moral or at least proto-moral in nature. In Chapters 4 and 5, we will argue that these adaptations cannot be reduced to either their component causes, such as genes and selection, or their input and output relations, that is, ordered sets of behaviours, responses, and reinforcements.

What makes the recurring structural features of the biological world that EMR calls natural moral kinds *moral* kinds? Why not agree with Street's view that moral responses are anchored in real features of the world that are not themselves moral in character? The point that we are working towards in this book is that the latter supposition is too empirically facile, if the kinds of things we are calling moral goods are in fact the sort of natural kinds that EMR supposes them to be. EMR opens up a promising line of empirical inquiry that would otherwise be closed.

Street's argument suggests in a loose empirical way that our moral responses will generally be linked to particular features of the world that it pays us to be morally interested in, but the generality Street's argument hints at is not the form of general explanation that EMR is after. Street is aiming at what she calls an "adaptive link" account: responses to particular features of the environment, like the screams of cats, turn out to be fitness enhancing in the human species and so they are selected for (James 2011, 184–185). EMR, on the other hand, promises to tell us what the regular features of the biological world are that would explain how we evolved to have the moral responses that we do, in a way that would explain the selective development of not only our own moral traits but also of similar traits in unrelated species in similar kinds of environments. The connection between the underlying structural features of these environments and the traits that developed in response to them will still of course be contingent, but it will be contingent based on what appear to be biologically necessary structural features of certain kinds of social environments.

Suppose that whenever we encounter things in our environment that have feathers, webbed feet, and a bill-shaped beak, we call such things "ducks." Biologists investigate these things, and they tell us that ducks turn out to be a species of water fowl. In this case, we would not be tempted to ask what makes ducks *ducks*. Similarly, the argument of this book seeks to establish two main things on EMR's behalf. First, there are interesting reasons to suppose that there are natural kinds of things in the biological world for our moral talk to be referring to. Second, there are interesting reasons to suppose that at least some of the time, when we are talking about morality, we are talking about these kinds of things. If EMR is right, we are in a position to know what morality ultimately is rather than to be deceived by the illusion of whatever we might have happened to have evolved to think that it is. If something waddles, quacks, and lays eggs like a duck, it is probably a duck. If EMR is right, something similar may be true when we say of something in front of us that it is morally good or bad.

What EMR makes possible is a truth-tracking account of morality tied to a general explanation for why we have at least some of the basic moral responses that we do (James 2011, 184). If EMR is right, the selection story regarding morality is at once much more complex, broader, and in the end more theoretically economical than the adaptive link account would suggest. As we understand it, EMR is committed to the claim that moral natural kinds are not historical accidents, like particular biological species, but more enduring biological kinds that are pervasive across a wide variety of species and processes of selection. EMR takes moral kinds to be contingent natural kinds, but it does not suppose them to be historically contingent kinds of things in the way that individual species are. The overall argument of the book is that it is plausible to suppose that such kinds exist, and that they are what we are ultimately talking and arguing about when we talk and argue about morality.

We think that part of what makes more philosophical arguments like Harman's and Street's seem empirically plausible is the current focus in evolutionary ethics on kin selection and reciprocal altruism, along with current accounts from comparative psychology of emotional responses in animals that seem to show positive reactions to situations that we humans would describe as involving fairness or empathetic caring. We believe that EMR makes empirically plausible a more unified and general argument for how such responses may have been selected for across a wide variety of species, including our own. Kin selection and reciprocal altruism can do some of the explanatory work but not all of it.

So, before we proceed any further with our discussion of EMR, we need to note that we are likely to find some version of Dawkins' (2006) selfish gene theory lurking behind other more standard efforts to link morality to evolution.

Does moral goodness reduce to what's good for our genes?

An affirmative answer to this question would immediately derail EMR. Biologists do certainly talk about a trait being good for inclusive fitness or good for an organism's genes, but this talk is metaphorical. Genes can build organic machines for which certain things are good or bad, like eating or being eaten. Such organisms can function well or badly, and their well-being, in this sense, can be promoted or damaged. But if some aspect of an organism is good for inclusive fitness, this just means its presence affects gene frequencies in ways such that certain genes or sequences of genes proliferate in a particular gene pool.

If natural moral goods affect trait selection, this may mean, mathematically, that certain genes do better in particular gene pools than certain other genes. But the evolutionary mechanism that is producing this mathematical result is the particular kind of trait in the particular kind of environment in which that trait exists. The functionality of a trait is not caused by the selection of the genes for it, but the genes for it are selected because of its functionality. This would be true for moral functionality as much as for any other kind.

Moreover, the moral natural kinds that EMR takes to be driving the selection of moral capacities are unlikely to reduce to lower level natural kinds. The regularly recurring pattern of what it is to help another as it exists in a corvid environment will be very different from how that same pattern might exist in a whale environment. While the underlying patterns may be the same, how these patterns are realized is likely to be different across different environments. A similar point is likely to be true for the capacities as well, as we will see in later chapters. For example, what Piaget calls instincts are capacities that he sees as specifically not being reducible to the behaviours that make them up, unlike reflexes. We take up the idea of moral instincts and how they might be related to EMR in Chapters 3–5.

Again, we think an important part of what makes an adaptive link account of morality seem plausible is the idea that behaviour patterns can evolve simply because of their relationship to inclusive fitness. Some behaviour patterns, the ones that we humans have come to call "moral," appear to favour the interests of organisms other than the organism so behaving. But this is not so: the behaviours in question turn out to favour what we might loosely refer to as the genetic interests of the organism exhibiting the behaviour. Such behaviour patterns will no doubt be triggered by particular features of the organism's environment, but these features of the environment are not of particular theoretical interest: what is theoretically interesting is the effect on inclusive fitness of the behaviour patterns in question.

One aim of this book is to break the hold that this paradigmatic way of thinking about evolutionary biology has had on our thinking about behaviours that appear to be moral or proto-moral. The argument of EMR is that from an evolutionary point of view, what is of primary importance in the biological appearance of morality is the occurrence of certain kinds of structural features of certain kinds of social environments. Instincts then arise and develop because of how they connect these features of the environment to behaviours that increase fitness. To understand this kind of trait selection, we need to understand the underlying environmental factors that are driving it. These arguments are developed further in Chapters 2–4.

Where do moral oughts come from?

According to EMR, moral goodness ultimately comes to rest with natural moral goods. Moral capacities have the deep structures that they do because of the deep structural features of the environment that they developed in response to. Monkeys can detect this kind of goodness in their environment, but they cannot name it or talk about it or recognize everything there is to recognize about it. We can name it and talk about it, but we too may be limited in our capacity to recognize all the important aspects of moral goodness. We can make mistakes about morality, both individually and collectively, in particular cases and in more systematic sorts of ways. And we are of course sometimes motivated to act by things other than morality.

Moral oughts, in contradistinction to natural moral goods, come from moral arguments and moral justifications. On most philosophical views regarding morality, morality itself begins with the human capacities for language, thought, and argument. If such philosophical views are naturalistically grounded, the argument is typically that while moral emotions make morality possible, they themselves do not create morality. So, while other animals may have similar sorts of emotions, these are not moral emotions, and hence, animals do not have morality. Animals may help each other, and they may do so because of an emotional response to the plight of the other, but there is no moral dimension to this emotion or the related behaviour without the ability to name and appreciate helping behaviour for the moral good it is or that we take it to be. Morality, on this sort of view, is a human artefact, based on our ability to reason about what we are going (or not going) to do.

In contrast, for EMR, morally good things arose before the human capacity for language and argument. Once we could name things and talk about them, we could name morally important things in our environment, like helping others, as well as morally bad things, like cheating. Once such things could be named and talked about, we could argue about their nature and how important they might be in this or that situation. Being able to talk and argue about moral values enabled us to form and defend moral judgments about particular situations, and this enabled us to look for and defend more general moral rules that might cut across a number of related situations and be formulated at various levels of generality. Being able to formulate a variety of general rules, along with more particular moral judgments, we would have also been able to combine together all our moral judgments into larger and larger sets of beliefs that were broadly coherent with one another.

Natural moral values, together with the capacity for language and argument, would have enabled the formation of what John Rawls (1971, 20–22) has called moral reflective equilibria. But EMR differs from Rawls over the interpretation of such equilibria: where Rawls (1980) argues for a constructivist as opposed to a realist interpretation of such equilibria, we argue for a combined interpretation. Many aspects of historical and geographically diverse equilibria will no doubt be amenable to a constructivist reading. Humans are richly inventive in terms of our social relationships and social rules. But as equilibria are developed across space and time, they may, in more or less significant ways, pull us closer to or further away from the natural moral values that lie at their base. In general, moral equilibria may track more or less closely the natural moral values that make them moral equilibria. The more closely they track natural moral values, the more closely they track moral truth; the less closely they track moral truth, the more likely they are to contain pockets of false judgments.

To return to the Canadian soldiers buried in Flanders fields, dying to protect one's group provides us with one kind of candidate for a natural moral value and a corresponding set of emotional or instinctual responses to that value. In the early years of the First World War, Canada had little trouble in attracting a sufficient number of volunteers to its army. As the war dragged on and its casualties grew along with questions about its purpose, the Canadian government of the day was forced to raise the idea of conscription and, finally, to institutionalize this idea in the last two years of the war. Both the idea and its institutionalization proved to be deeply divisive in the public arguments surrounding Canada's war effort. Those who favoured conscription called those who opposed it disloyal cowards. Their opponents tied the war to the nationalistic values and imperialistic aims of Britain and the other European powers Britain was fighting with, values and aims that arguably should have little or nothing to do with Canada. When, where, how, and why in-group loyalty is appropriate are interesting and important moral questions.

Much of the argumentation around such questions will be directed at particular social institutions in particular social circumstances. But in the conscription case, one of the two groups may have been more morally correct than the other: perhaps either more or less in-group loyalty was the morally right course of action. So, even as Canadians continue to buy their poppies every year to remember those who died on their behalf, they might do well to remember and to reflect on something else:

> In the late nineteenth century and the first decade of the twentieth, nothing reshaped the world more than European imperialism. It redrew the map, enriched Europe, and left millions of Africans and Asians dead.
>
> (Hochschild 2018, 150)

It also left Europe itself on the brink of war. The Paris peace process of 1919 did little to dampen either European nationalism or imperialism, and so we might wonder what might have happened if the German war effort had not collapsed as it did. Probably not a League of Nations, but even so, it may perhaps have been that those who opposed greater Canadian participation in the Allied war effort were morally right to do so.

Our example here is intended to be open-ended and provocative. It is not meant to be flip: we fully recognize that the questions we are using it to raise are going to be both morally and historically much more complicated than our short quote might suggest. We take up the general question of the limits of reasonable forms of moral partiality in Chapter 7, having in Chapter 6 taken a more careful look at a more fully developed historical example of moral argument and moral change: the nineteenth-century British abolition of the trans-Atlantic slave trade and the abolition of slavery itself within the United States. As we develop it, EMR will understand moral ought-claims as propositional claims arising from moral justifications embedded in moral reflective equilibria.

Moral oughts will thus occur in statements or judgments that are the conclusions of moral arguments, where the premises of such arguments are other moral statements or judgments from the moral reflective equilibrium in question. Entailment relationships always exist within the context of a moral reflective equilibrium, argumentatively linking moral judgments to other moral judgments. According to EMR, the more closely the equilibrium tracks natural moral values, the more likely it is that the justificatory arguments made on the basis of this equilibrium are ultimately sound. Moral arguments eventually come to an end, and then we are left with questions over the truth of the premises of such arguments.

If we suppose that the truth of a set of judgments in a moral reflective equilibrium ultimately depends on how closely that equilibrium tracks natural moral values, we are not thereby committed to any entailment relationships that would take us from natural (moral) facts to moral oughts. EMR does not commit the inferential fallacy of deriving ought-claims from is-claims. Rather, it connects natural moral values to the ultimate soundness of moral arguments, not to their validity. It does not deductively take us from "is" premises to "ought" conclusions.

To the deeper normative question of "why ought I be moral?" EMR gives the following answer. To be posed, this question requires language and argument. The general point of moral argument is to determine jointly and reasonably the morally best course of action in cases that are morally contentious. If one asks, at the end of such an argument, why one should proceed with the best course of action, so determined, one is either doubting the validity of the particular process that led to the particular result in question or doubting the validity of the general process itself. If it is the latter, the question then becomes why one was engaged in such a process to begin with – perhaps to fool others. But to the degree that we are aware of and genuinely moved by natural moral values, as well as by the moral arguments that arise in response to these values, we have reasons to be moral. Sometimes it may be right to fool others, sometimes not. It depends on the details of the reflective equilibrium in which such particular judgments about when and whom to fool are embedded.

The structure of the book

The earlier chapters of the book make the argument that some of the basic things we refer to as morally good can plausibly be identified as an enduring biological kind or cluster of such kinds. Later chapters recognize that human societies involve complicated institutional arrangements that include an enormous variety of diverse roles, responsibilities, and rules and regulations. Many of our arguments about how we ought to act will be internal to our shared social understandings of these institutional arrangements. Even so, if these institutional arrangements can also be shown to have grown out of and still be framed by our human understanding of the natural values discussed in the first part of the book, we have plausible reason to suppose that these natural values are moral values. This does not commit EMR to the claim that all moral values are natural moral values, or that all moral values can be directly traced back to natural moral values. But if natural moral values can be shown to be at the foundation of human moral systems, there is an independent form of ontological truth for these systems to be tracking.

In Chapters 2–4, we begin to lay out the empirical case for EMR. Like the adaptive link account of morality, EMR is meant to be an empirical hypothesis regarding the evolutionary origins of morality. Unlike the adaptive link account, EMR takes moral properties to be a particular kind of natural property, a natural kind property to be found in regularly recurring structural features of certain kinds of social environments. Given the current empirical evidence regarding the evolutionary development of such responses, we think EMR is a plausible alternative to evolutionary accounts of ethics that, like the adaptive link account, take moral properties to be essentially the product of human reason and human emotions.

The general idea of Chapter 2 is that what EMR identifies as natural moral values arise much earlier in evolution than other theories would have moral or proto-moral responses arise in humans or other species. If EMR is right, natural moral kinds are a pervasive biological phenomenon, deeply embedded in the structural features of the environments of a diverse range of species, some of them more or less closely related to one another but others much more distantly related.

With the evolution of humans, and language, thought, and reason, something morally special does indeed happen. According to EMR, what happens is that humans became capable of naming, talking, and arguing about the moral values that were already a significant part of their social environment. Once we were able to argue about these values, we were also in a position to institutionalize them in a wide variety of ways, creating along the way diverse arrays of derivative moral values of a more essentially cultural kind. EMR does not deny that there are significant constructive and cultural elements to human morality, but it does claim that these elements of morality are ultimately grounded in the natural kinds of moral values that we discuss in Chapters 2–4. Moral arguments depend on consistency and coherence for their internal validity, but what makes them ultimately sound are the natural moral values that began the cultural practices of morality among humans and that continue to frame these practices within morality as an enduring biological reality.

While Chapter 2 looks for the earliest beginnings of natural moral values in evolutionary development, Chapter 3 begins to trace out how such values might be expected to have become more numerous and more complex as integral parts of social environments that were themselves becoming more complex and more dynamically interactive. Chapter 3 also begins to make the key empirical argument of EMR that the structure of moral responses is patterned by more widely occurring structural features of the social environments in which these traits are selected for. Both chapters draw heavily on empirical evidence that is rapidly accumulating in evolutionary and developmental psychology. What is new about EMR, and what we think is empirically interesting and important about EMR, is how it theoretically interprets and synthesizes this evidence.

Chapter 4 more directly considers the empirical question of whether what is ultimately real about morality is to be found exclusively in the psychological features of certain kinds of organisms, or whether it should also include the kinds of structural features of the environments of these organisms that EMR is calling natural moral values. We are not supposing this to be a readily solved question, and we are certainly not proposing to have solved it here. What we are arguing, however, is that it is not as simply resolved as one might think, by simply appealing to Occam's razor in favour of the idea that all that evolves are behaviour patterns and any psychological capacities that might better facilitate them. As traits of organisms, this is all that evolves, but the idea motivating EMR is that there are commonly occurring structural features of certain kinds of environments that are causally involved in the selection of these sorts of traits. Whether this is so is ultimately a question of explanatory power and adequacy, as well as simplicity. What we are explicitly arguing here is that it is worthwhile to look in the direction of EMR for all three sorts of reasons.

EMR's key hypothesis is that the structural features of moral environments are causally responsible for the selection of moral traits in organisms, whether these be purely behavioural, instinctual, emotional, or cognitive. EMR supposes that these traits, as they develop, may themselves help to structure moral environments, particularly as the complexity of such environments develops over time. This process amounts to a moral version of what biologists call niche construction or causal feedback between organisms and their environments. EMR thus sees psychological traits with moral foci as part of the evolutionary explanation for the existence of morality, but it takes such traits to be, by themselves, an incomplete account of morality's existence in the natural world of evolutionary biology.

Connecting the natural values of the earlier chapters of the book to moral reasoning and moral justification is of central importance to the overall argument of the book. If the earlier chapters make plausible a moral ontology of natural moral values, it is the task of the later chapters to make plausible the link between these values and moral justification.

With Chapter 5, the book turns more explicitly to the connection between claims about the way the biological world is and claims about the way the moral world ought to be. It has been argued philosophically that moral values may reduce to something other than the evolved psychological features of species like humans, whether in the natural world or in some sort of unnatural world of the kind inhabited by numbers, gods, or the universal laws of logic. Chapter 5 covers some of the main naturalistic arguments in this vein. Although we think that these approaches contribute to explaining the normativity of human morality, they do not include all the moral content provided by the sorts of psychological traits discussed in Chapter 4, which EMR takes to be a crucial part of morality, nor do they include all the moral content provided by the natural moral values that we think may be structuring these traits. Like moral sense theories, the naturalistic approaches to morality we consider in Chapter 5 contribute to our understanding of morality, but they do not provide a complete account of morality.

Chapter 6 considers a human case where a historically important change in moral beliefs may have been due at least in part to unlooked-for moral values in the social environment in which that change occurred. Chapter 6 runs parallel to Chapter 4 in that both chapters seek to reveal certain kinds of structural features in the environment as causal sources for certain kinds of moral responses. The main difference in Chapter 6 is that we are now talking specifically about human moral beliefs as the kind of moral response in question, as opposed to the allegedly proto-moral responses of other species. This is a key part of our argument that the natural kinds we discuss in the first part of the book are moral kinds and also that the responses of all species to these kinds are moral responses, if they are aimed at the kinds of things in question in an appropriate way.

Chapter 7 returns us to the problem raised at the outset of this first chapter. If morality has biological roots, should we expect morality to be inescapably tribal, dividing the social world into those individuals who count, from a moral point of view, and those who do not count or count less? We briefly wondered about the source of the obligation to continue a war that others were fighting in, supposing that the source of this obligation is to be found in the moral importance of loyalty to the groups that we depend on and are parts of. Chapter 7 asks how exactly EMR may affect our understanding of the relative weights of partial and impartial moral reasons in our moral thinking as humans. We think that EMR is a morally interesting theory that may help us better understand important questions in the philosophical and practical study of morality.

Drawing on earlier chapters, Chapter 8 provides a more detailed discussion of what is probably the main philosophical question facing EMR as an empirical theory of morality. How do the natural is-claims going in at the one end of the theory get transformed into the moral oughts allegedly coming out the other end of the theory? How can empirical considerations about what is the case lead to moral conclusions about what ought to be the case? Can we really get from the first half of the book to its second half? It may seem as if EMR is caught in an impossible dilemma: either the values at its heart are entirely natural, and hence there is no real reason to suppose they are moral, or these values can be connected to moral ought-claims, but only in ways that involve some form of invalid inference from is-claims to ought-claims. The final chapter of the book is devoted to the argument that EMR can in fact slip between the horns of this dilemma.

Bibliography

Bartal, Inbal Ben-Ami, Jean Decety, and Peggy Mason. 2011. "Empathy and Pro-Social Behaviour in Rats." *Science* 334:1427–1430.

Brosnan, S.F., and Frans B.M. de Waal. 2003. "Monkeys Reject Unequal Pay." *Nature* 25 (18):297–299.

Collier, John, and Michael Stingl. 2013. "Evolutionary Moral Realism." *Biological Theory* 7:218–226.

Dawkins, Richard. 2006. *The Selfish Gene (30 Anniversary Edition)*. New York: Oxford University Press.

Hamlin, J. Kiley, and Karen Wynn. 2011. "Young Infants Prefer Prosocial to Antisocial Others." *Cognitive Development* 26 (1):30–39.

Harman, Gilbert. 1977. *The Nature of Morality: An Introduction to Ethics*. New York: Oxford University Press.

Hochschild, Adam. 2018. "Stranger in Strange Lands: Joseph Conrad and the Dawn of Civilization." *Foreign Affairs* 97 (2):150–155.

James, Scott M. 2011. *An Introduction to Evolutionary Ethics*. Chichester: Wiley-Balckwell.

Moore, G.E. 1903. *Principia Ethica*. Cambridge: Cambridge University Press.

Pellis, Sergio M., Vivien C. Pellis, and C.J. Reinhart. 2010. "The Evolution of Social Play." In *Formative Experiences: The Interaction of Caregiving, Culture and Developmental*

Psychobiology, edited by C. Worthman, P. Plotsky, Schechter D. and C. Cummings. Cambridge: Cambridge University Press.

Rawls, John. 1971. *A Theory of Justice*. Cambridge, MA: Harvard University Press.

Rawls, John. 1980. "Kantian Constructivism in Moral Theory." *The Journal of Philosophy* 77 (9):515–572.

Street, Sharon. 2006. "A Darwinian Dilemma for Realist Theories of Value." *Philosophical Studies* 127 (1):109–166.

Warneken, Felix, Brian Hare, Alida P. Melis, Daniel Hanus, and Michael Tomasello. 2007. "Spontaneous Altruism by Chimpanzees and Young Children." *PLoS Biology* 5 (7):e184. https://doi.org/10.1371/journal.pbio.0050184.

2 The moon in the water[1]

Evolution and moral realism

In a Japanese woodblock print by Koson, two monkeys hang down from the branch of a tree at dusk (see frontispiece, p. iii). They are above a pool of water, one monkey hanging from the hand of the other. The lower monkey, closer to the pool of water, is reaching out to touch the moon that glows luminously below it in the water. According to EMR, moral goods are like the moon in the water: the reflection of a real part of our natural environment as the particular kind of social and intelligent primate species that we are. Different species of organisms may be looking for and at moral values through different kinds of response mechanisms, but what they are responding to through these mechanisms is something that is really there.

On most other evolutionary approaches to ethics, perceived moral values are an illusion reflecting no independent moral reality. Proponents of one general line of approach, which we might label the standard view of evolution and ethics, range from biologists like Dawkins (2006), Wilson (1998) and Alexander (1987), to psychologists like Rachlin (2000) and Hauser (2006), to philosophers like Ruse (1986), Joyce (2001), and Nichols (2004). According to this view, certain cooperative behavioural patterns develop and so become biologically real, but morality itself doesn't become possible until creatures like us evolve a sophisticated enough cognitive ability to mistake the more immediately apparent goals of such behavioural patterns for (apparently) objective moral values. Ethics is possible for humans because of our evolutionary heritage as cooperative primates, but it only begins after we appear on the biological scene and start to talk, to argue, and to reason about how we ought to live our lives together in larger and larger cooperative groups. On this line of thought, the best theory of moral belief is one that treats all such beliefs as a particular kind of cognitive error that makes ethics normatively efficacious for humans (Mackie 1977, 42–49). If we were not prone to thinking that moral values were real, we would not pay enough attention to them enough of the time to survive and flourish as we do.

Along with the standard view, we think morality is tied to cooperative behavioural patterns. Against this view, we think that moral values are a real part of the biological world, whether or not animals are able to perceive them. Moral values arise as particular kinds of good things, the pursuit of which serves to enhance particular kinds of cooperative behavioural patterns. Particular moral goods are moral goods because that is simply the kind of good thing that they are. At its empirical best, the error theory of moral belief is an appeal to the simplest explanation for why we humans are as concerned about morality as we appear to be. EMR's response to the error theory is to offer the beginnings of a more general explanation that is based in the evolution of animals in contexts where moral values are adaptive.

EMR hypothesizes that being able to recognize and pursue moral goodness is good for the survival of intelligent social animals. But being good for survival is not the same thing as being morally good. "Good for survival of x" is simply short for "enhances fitness of x in such and such an environment." Moral values, though, are just that: certain kinds of good things that ought to be pursued or certain kinds of bad things that ought to be avoided. At this first level of explanation, we are tying these "oughts" to biological imperatives attached to positive and negative moral values. The fact that moral values are good for survival is the explanation, we think, of their existence but not their normative force. The normative force of moral values is originally to be found in the morally good things themselves, in their appearance in environments as things to be pursued.

As language developed in the human species, various kinds of norms were articulated, argued about, and agreed upon through social and cultural evolution. Cultural norms include, significantly, moral norms. We do not mean to suggest that the normative force of these moral norms is fully exhausted by the biological imperatives attached to the natural moral goods to which some moral terms may ultimately refer. What we do mean to suggest is that at the very beginning of moral discourse, before there were moral terms, there were already morally good things to be referred to by such terms. In short, we are thinking of terms like "empathetic caring" as rigid designators, naming particular natural kinds, which we are identifying as moral goods.

What makes such to-be-pursued things moral is that they are a particular kind of good thing, a good kind of thing that humans have come to recognize and to talk about as morally good. Because moral goods are a particularly salient kind of environmental good, moral normative force is rooted in a particularly strong kind of biological imperative. But however strong such biological moral imperatives might be, our more immediate point here is that there is more to moral goodness, as we have just described it, than what might morally matter to particular animals or even to particular species of animals. Moral values may exist independently of any particular species' ability to detect them or to be motivated by the biological imperatives to which they are organically connected.

According to EMR, moral values arise as important parts of the social environments in which some species of animals survive, reproduce, and evolve. Moral values are not unlike predator–prey relationships. Just as animals need to be able to recognize predators and be moved to act on this recognition, social and intelligent animals need to be able to recognize (say) fairness and to be moved to respond to it in morally appropriate ways. Like predator–prey relationships, natural moral values are likely to be significant for but not reducible to the fitness of the kinds of animals for which they matter. Although animals of these kinds might sometimes err about the moral values that are part of their environments, they cannot err much of the time, never mind all of the time. Similarly, detection of moral values is unlikely to be normatively inert because unless the capacity to respond to moral values is at the same time a normative capacity (possibly teleonomic but not teleological), there is no good reason for it to evolve. Where explicit recognition of moral values and motivation explicitly by them are not within the capacity of the creatures involved, such creatures might still be moved by moral values if that increases their fitness. Like predators, moral values are extremely important parts of an animal's environment. Ignoring them could be perilous.

The evolutionary origins of moral goodness

If morality is central to human existence because of the kind of evolved animals we are, we are unlikely to get it entirely wrong. We can, of course, make mistakes, and some of these mistakes will no doubt be significant. Stingl (1996) and Stingl and Collier (2005) discuss several systemic kinds of moral mistakes human moral thinking may be prone to. Moreover, following Atran and Norenzayan (2004) and our commentary in Stingl and Collier (2004), some moral values, like many of those tied directly to religious traditions, may be illusory in exactly the sense intended by the error theory of morality. On the other hand, common sense morality, as well as the general moral theories that respond to it, like Kantian ethics or utilitarianism, is likely to get many important things right.

Complicating matters is the fact that religious values and natural moral values are often intertwined in human cultural responses to their environments in ways that make these different kinds of values difficult to pry apart. While religious beliefs may always have an illusory component, some religious beliefs may also track real moral values. In terms of punishment and cooperation, and in terms of the apparent externality of moral obligations and prescriptions, the relationship between religious and moral beliefs is as important as it is complex. It is an issue that needs to be addressed by a view like ours, although we are unable to address it at this point in our development of EMR.

Returning to common sense morality, what sorts of natural moral goods are we liable to get right? Here is a rough list – empathetically caring about and doing what we can to ameliorate the pain of others; sharing pleasures as well as sharing pains when there is nothing else to be done; sharing food and other material resources, helping others when they need it; longer term cooperative relationships where we work together in mutual support and for mutual benefit; the forms of trust and loyalty that enable and enhance such relationships; reconciliation when such trust has been breached; caring about common goods, as well as the individual goods of those most immediately socially connected to us; fair treatment of others, including, within certain bounds, the punishment of cheaters. We are starting with the assumption, not that all these moral values are in fact natural moral values, but rather that if we are to begin to look for natural moral values, common sense morality is a good place to start.

We should also take note that all these common sense moral goods can be used to undercut one another. Trust, loyalty, and longer term cooperative relationships can all be used towards ends that are ultimately inimical to these moral values themselves or to other moral values. Empathy can be used to manipulate others effectively or, in the extreme, to torture them in the most soul-destroying ways humanly imaginable. Punishment can be used to control and dominate. If positive moral values are at the centre of human existence, their perversion may likewise be at the centre of moral evil. In evil actions, moral values are used in ways that are inimical, in the long run, to the well-functioning of the underlying values. In this book, we do not address moral evil but we recognize its possibility as a very real evolutionary correlate of natural moral values.

In addition to positive moral values, that is, moral goods, there are things that are morally bad. Cheating and cruelty, for example, arise alongside trust and empathy. It may be that morally bad things are parasitic on morally good kinds of things arising in an environment and having reproductive value. An organism cannot cheat, for example, without a system of mutual cooperation to exploit. While questions about moral evil and negative moral values more generally are important, in this book we focus on positive moral values as our paradigm of what counts as moral natural kinds. The point of the book is to begin to develop a research program that treats moral values at their origin as natural moral kinds, and to carry out this task we begin with positive moral values.

To begin at the beginning, EMR sees moral goods as natural products of the evolutionary processes that create intelligent social beings. As biological organisms evolve, certain kinds of things become good for them. Different kinds of things become good for different kinds of organisms in different ways. Many if not all of these good things will be good for the inclusive fitness of the organisms involved, but on our view this kind of reproductive goodness is not the causally efficacious sort of goodness that is involved in the actual selective processes themselves.

Consider nutritional goodness. Nutritional goods are good for inclusive fitness, but this is not what makes them nutritionally good. That is, their link to inclusive fitness does not make nutritional goods the particular kind of good thing that they are. What makes nutritional goods nutritionally good is their role in the functioning of organisms, the conversion of food energy to bodily energy. In biological terms, we understand nutrition as a function of development or ontogeny, not evolution or phylogeny. Nutritional goods are good, in a general way, for all organisms. Moral goods are not good for all organisms, according to EMR, but only for a particular class of organisms: those that are intelligent and social, or are at least able to be positively affected by moral goods in their environment.

Capacities that respond to natural moral goods may come under distinct evolutionary mechanisms and forces and form an evolutionary branch, not in species evolution but in trait evolution. We call this phenomenon an evolutionary trajectory, and we explore these ideas in more detail in the next two chapters. There is no implication of any endpoint or purpose in our use of the term "trajectory," but starting on an evolutionary trajectory can permit further enhancement of the trait involved, even if, as in the case we are interested in here, moral goods are not directly recognized at any point in the trajectory. At the beginning of a trajectory, things can be good to eat before organisms develop the capacity to recognize this particular form of goodness as it exists in their particular environment. So too with moral goods: they can exist as the particular kind of good thing that they are before the organisms involved develop the capacity to recognize them for what they are. Consider some of the simplest entries on our list of moral goods: empathy, trust, and fairness. It is doubtful, for example, that capuchin monkeys recognize fairness as the good kind of thing that it is for them; yet it is undoubtedly a good for them, whether they recognize it as good or not.

In Brosnan and de Waal (2003), capuchin monkeys who got cucumbers, while others got more highly valued grapes for performing exactly the same task, typically refused their lower valued reward. This experiment, and Brosnan and de Waal's interpretation of its result, has spawned a small literature (see Fletcher (2008) and van Wolkenten, Brosnan, and de Waal (2007) for a review of this literature). The subsequent experiments of Fletcher (2008) and van Wolkenten, Brosnan, and de Waal (2007) seem to confirm Brosnan and de Waal's 2003 finding of an aversive response to inequality, ruling out other explanations such as frustration or a more simple and straightforward interest in the more highly valued reward (I don't want this, I want that). Although it may always be difficult to say much about the actual cognitive content of capuchins' emotional response to inequality, it does seem clear that at some level they are able to recognize the presence or absence of fairness when it comes to matching efforts and rewards, and that they care deeply about whatever it is that they recognize about such situations.

Knowing what fairness is and being able to recognize it in some way or other may thus be two (or more) very different things, and so capuchins may be able to recognize fairness without being able to recognize this particular kind of thing for what it is. What interests us here is *that* capuchins seem to recognize fairness, not *how* they recognize it. Exactly what sort of evidence might indicate that a particular species recognizes fairness as fairness is another question for another day. Experiments with dogs (Range et al. 2009) may serve to emphasize the point we are after, because they too cooperate less in the presence of unfair rewards. In the case of dogs, however, where the dog not getting a treat stops shaking its paw with the experimenter faster than it otherwise might, one of the alternative interpretations of such results may prove more robust, such as the fact that there is an edible treat in the environment of paw shaking that is going past the nose but not into the mouth. That the treat goes into a nearby mouth (the other dog's, which is both shaking paw and getting a treat) may be beside the point, at least at the level of dog cognition. It may even be that for dogs long domesticated, loyalty or trust may interfere with the aversion to unfairness: unlike dogs, wolves are much quicker to stop cooperating when rewarded unequally (Essler, Marshall-Pescini, and Range 2017). Alternatively, the artificial selection of domestication might have dumbed dogs down. Or it might have made them more hierarchical. Or both. What nonetheless seems clear from the cases of monkeys and dogs is that fairness can matter to social and intelligent creatures long before they are able to recognize fairness itself as an important feature of their social world.

For our argument here, questions about what it is that capuchins or dogs recognize about unfair situations are interesting but not immediately important. Such animals recognize, at some cognitive level or other, a pattern in their environment that we are also able to recognize, a pattern that we refer to as being unfair. This thing that we refer to, unfairness, is what EMR regards as a natural moral value. Fairness matters to the overall health and well-being of social and intelligent mammals, and at a particular point in the evolution of cognitive development, animals evolve who can clearly recognize this for themselves. Along this trajectory of evolutionary development, recognitional capacities can vary, but the better a species is able to recognize fairness, the better its members will be at performing patterns of behaviour that are better for them biologically as well as morally.

Our key point here is that fairness can become an important feature of a species' environment long before its members can demonstrate an emotional aversion to unfair situations or behaviours, whatever such a psychological state might amount to or be directly aimed at. Nutritional goods arise before organisms are able to aim at or track them. But as organisms evolve that can track particular kinds of nutritional goods, some may be better able to track these goods than others, and an evolutionary arms race can develop. Tracking mechanisms may thus improve. Depending on ecological factors, these mechanisms may aim more or less directly at the particular kind of good that they track. Likewise, for social and intelligent species, certain kinds of things can become morally valuable, and given ecological pressures, species may come to track these moral goods more or less directly. We are able to track fairness more directly than capuchins, and they may be able to track it more directly than dogs. The point is that all three species seem to be tracking the same general kind of moral good, namely fairness.

Pistol shrimp

To push the argument further, let us consider a particular form of unfairness. Deceptive behaviour is a form of unfair behaviour, and it can be expected to arise wherever cooperative behaviour patterns arise. But we need to proceed carefully here: terms like "deceit" can have both moral content and merely descriptive content. Descriptively, deceit involves producing a misleading sign that results in an advantage to the organism producing it. Morally, deceit involves producing a false signal in order to produce such an advantage. In this sense, deception is certainly unfair: it exploits a cooperative signalling system for individual advantage. The problem is that in applying terms such as "deceit" to behavioural patterns, descriptive content does not immediately entail moral content.

Consider signalling behaviour in male big-clawed snapping shrimp (*Alpheus heterochaelis*), also known as pistol shrimp. This is a colonial, monogamous species. Male shrimp fight over resources, and their ability to win fights is determined by their relative body size. Body size is highly correlated with the size of a shrimp's claw, and rather than estimate body size by wrestling with one another, shrimp first signal their size to one another by opening and closing their claws, shooting out a pulse of water (Hughes 1996). The signal is detected by mechanoreceptors on the claw of the other shrimp by way of the water currents thus created (Herberholz and Schmitz 1998). Some shrimp have claws, however, that suggest their body size is bigger than it really is. In encounters with shrimp that are slightly bigger, shrimp with deceptively sized claws open and close them more frequently, resulting in an increased signal at the mechanoreceptors. Honest signalling systems may thus create the evolutionary opportunity for deceptive signalling long before the organisms involved can recognize deception for what it is or even reliably detect it. If the light is good and the water clear as in a laboratory environment, shrimp deception seems to be detectable through escalation to wrestling, suggesting an alternative visual channel for size detection (there is also a chemical channel that signals past experience in fights). Matters are less clear in the shrimp's natural environment, where the light is poor and the water cloudy and turbid.

On the surface, the agonistic behaviour of pistol shrimp has many of the features of typical biological discussions of morality: cooperation (they engage in ritual displays), altruism (they do not kill opponents), cheating, and detection (exaggerated signals and alternative channels for detection). Things are not so simple, however. The evolution of pistol shrimp agonistic behaviour is not currently known, but we can speculate. Unlike many cases of fighting organs, the large claws of pistol shrimp probably did not evolve for intraspecific fighting but for predation, so their existence can be explained by individual selection for nutritional advantage. It is then advantageous for them to be used in fighting for territory. The evolution of sensors to detect claw size, and hence fighting ability, has a clear advantage to individual shrimp, similar to the advantage that comes from being able to detect fighting experience of potential adversaries through chemical sensors. It could be that the failure of shrimp to kill their opponents is a consequence of these other fight-limiting processes and the balance of risk of fighting versus potential damage, making fighting to the death of little advantage. Deceit by overactive large-clawed individuals has individual advantages, like most forms of deceit, and there is no evidence of punishment of deceit. At best, there is evidence that the advantages of deceit are limited by other channels for detection of fighting ability. So deceit and deceit detection do not in this case seem to have any but individual advantages. Perhaps deceit is too costly to become widespread in comparison to other strategies, but it is useful enough to open a niche for itself if most of the population is not deceitful.

If this individual selection story is true, it isn't clear that the behaviour of the large-clawed shrimp is in any sense unfair. There is nothing resembling a coordinated signal system with any common origin. The "signalling system" isn't really a system at all. Nobody is exploiting a system of mutual cooperation to produce an advantage only available to them through their disregard of the goods of others.

On the other hand, pistol shrimp are colonial, and perhaps colonies that limit fighting are more prosperous than ones that do not. If so, there may be some element of group selection present, and in this case, some general form of harm reduction may enter into the evolutionary process. Perhaps the evolutionary story is some combination of the two selective mechanisms. Without further investigation, we cannot tell. The behaviour would be the same in either case. Without further investigation into evolutionary history, or perhaps the internal causes of the identical behaviours, we cannot tell if group selection is required, or in fact occurred, at some point in the evolution of this particular species. The best we can say is that invasion by "killer shrimp" is unlikely now due to individual disadvantages. On the issue of deceit, moreover, the deceit involved is energetically costly, and perhaps colonies with too much deceit are not as prosperous as ones with limited deceit. Again, group selection cannot be ruled out. The evidence we have is ambiguous between self-interest and broader interests as the mechanism(s) of selection. Moral (or proto-moral) properties of the environment may or may not play a role in the selection history.

So things that look like moral goods (or bads) may not be such things at all. On the other hand, they might be, and thus they might exist, long before an organism has the more sophisticated cognitive capacities of a dog or a capuchin monkey to recognize them as important parts of its environment.

Natural moral values and evolutionary trajectories

The shrimp example should make us wary of moving too quickly from observed behavioural patterns to assumptions about underlying moral values' causal involvement in such behaviour. Some forms of cooperation, for example, may turn out to be moral dead ends, with no possibility of further evolution into more fully moral behaviour. In the shrimp case, the cooperation might be so weak as to really be just coordinated behaviour with no other-related interests playing a causal role in that behaviour.

To focus more specifically on the causal role of other-related interests in moral behaviour, we move to a more familiar example, cooperative behaviour among bees. In some species of bees, female worker bees feed male and female larvae at a ratio of 1:3, parallel to the degrees of genetic relatedness involved (they are related three times as closely to the female larvae as to the male larvae). This aspect of larval feeding does not, apparently, involve cuing to the individual goods of the larvae they feed. When enslavement occurs among ants that also exhibit the 1:3 feeding pattern, if there is no genetic relationship between enslaved female workers and larvae, the ratio of feeding becomes 1:1 (Gould 1977, 264–265). What this suggests is that worker insects recognize, probably chemically, particular larvae, as well as the sex difference between these larvae. What they are not recognizing or responding to are the individual goods of the larvae to which they are providing more food. Their form of cooperation, in other words, is not causally rooted in the good, nutritional, or moral, of the other, even though their feeding behaviour does nutritionally benefit the larvae they are feeding as a consequence of that behaviour.

What might happen were bees to evolve a sufficient level of intelligence to recognize the nutritional goods of other bees in the sort of case we describe?

Could they thereby be motivated to respond to these goods, as such? Probably not, given the strong degree of relatedness of female bees to female larvae along with other aspects of bee reproduction. Given the strong form of kin selection that is behind their form of cooperation, theirs may not be the kind of cooperation that could ever be causally tied in the right way to the good of another. There are other aspects of bee larvae feeding behaviour that might more easily cue to the needs of the larvae. These aspects can be very complex; intelligence might help to further develop such behavioural patterns, but it might just as well hinder them, for all we know. In any case, this sort of cuing could be compatible with our account of the evolution of morality, because it could be directed to the good of another. The singular thing about social bees, ants, and termites is that they are like superorganisms (Holldobler and Wilson 2008), and in this respect are very different from most organisms. Inasmuch as they are superorganisms, their behaviour cannot be properly understood at the individual level only (though individual behaviour is not irrelevant either).

We thus have come to doubt Darwin's (1874, 99–100) point about bees and morality. Darwin suggested that were bees to become intelligent, they would no doubt have a completely different morality from ours, based on the differences between their cooperative behaviour and ours. Where we might face a fairly straightforward moral imperative to feed our siblings when they are hungry, the bees, presumably, would face at least two imperatives, along the lines of "Stuff your sisters" and "Don't worry too much about stuffing your brothers." But why suppose these latter two imperatives are moral imperatives? Like the form of cooperation these imperatives would supposedly develop from, the intellectual capacity capable of producing such imperatives is unlikely to be on a possible trajectory of moral evolution. Bees don't start from the right sort of place, if we are right about the kind of thing that moral values are. These kinds of things seem unlikely to arise in bee environments, given the way their form of cooperation works for them.

From this discussion it might seem that if there are environmental conditions for morality (arguably lacking in the bee case) and there is adaptation to these conditions (where the conditions play a causal role in producing the adaptation), then the resulting adaptation will be moral. This is undermined by the pistol shrimp example, however, because the existence of sociality (colonies) allows for group selection but does not imply it, even though sociality is obviously required for agonistic behaviour to arise. The same behaviour can be explained in terms of individual or group selection. Unlike with the bees, both possibilities are on the table. The sociality (pistol shrimp interact with each other within colonies) plays a role in the individual selection story, but it does not imply any other-directed interests. The existence of an adaptation to moral goods in the environment no more implies any sort of moral tracking of those goods than an adaptation to a nutritional good in the environment would necessarily imply that the adaptation is nutritional. The organism must make use of the nutritional resource in a nutritional way for it to be a nutritional adaptation or at least create the potential for the resource to be used nutritionally.

Similarly, light makes vision a possible adaptation, but not all adaptations to light, such as change of skin colour, are visual adaptations. There are many features of the world that can be recognized by optical means, but some means are better than others. Some of the means, rudimentary or sophisticated, will be on developmental pathways that make it possible to get from more rudimentary forms to more sophisticated ones through evolutionary processes. Some rudimentary forms of light detection will not be on any such pathway, and hence, will not be any form whatsoever of optical capacity. For example, skin cells might respond to light by darkening, but even though it is an adaptation to light, this would not be on an optical trajectory if the capacity permits no differential sensitivity to the light that is present at a particular time. Light sensitivity can only evolve into optical detection if it allows differences in light intensity and/or hue to be detected.

Moral capacities

So how might moral capacities be like nutritional or optical capacities in this same important way? Consider this experiment with capuchin monkeys summarized in de Waal (1989, 104):

> Several monkeys were trained to pull chains for food. After they had learned this response, another monkey was placed in an adjacent cage; pulling the chain now also caused the neighbour to receive an electric shock. Rather than pulling and obtaining the food reward, most monkeys stopped doing so in sight of their mate's suffering. Some of them went so far as to starve themselves for five days. The investigators noted that this sacrifice was more likely in individuals who had themselves once been in the other monkey's unfortunate position.

What is going on in the heads of the capuchins that refuse to pull the chain? One thing that might be going on is this: if you set two people down next to one another and prick the finger of one of them while the other watches, there are two kinds of brain responses (Singer et al. 2004). In the first person, some parts of the brain register the sensory aspects of pain, including such things as its location and intensity. Other parts of the brain register the affective aspects of pain, such as the subjective experience of its unpleasantness. In the second person, there is no activity in the sensory part of the brain but the same kind of activity in the affective part of the brain. This second kind of response appears to be automatic, and it appears to be a means of allowing the second person to feel the affective aspects of the pain of the first.

This sort of automatic response to the pain of another may be the first and most rudimentary version of what Nagel (1986) calls the view from nowhere: the bare beginnings of an impartial point of view from which we register the negative and positive aspects of things like pains and pleasures without registering whose pains and pleasures they are. From the view from nowhere, we observe only that pain exists and that it is bad. Not that our pain is bad, or that the pain of the other is bad, but simply that pain exists and that this is bad. On this view about rudimentary forms of pain recognition, the organisms involved don't need to know, in any sense, who is who, or whose pain is whose: they just need to register that pain is occurring and that this is a bad thing: the sort of thing that by its very presence demands amelioration. While we do not think pain itself is a natural moral value, presumably a bad one, we do think that responding to and ameliorating the pain of another may be a morally good thing.

This may be what is happening for the capuchins in the above experiment. This claim can be empirically tested, so it provides one important kind of test for the theory we are developing here. On this theory, the moral capacity of the capuchins enables them to feel the pain of another as something that needs to be ameliorated. They automatically experience the pain of the other, not their own pain, and not in a way that the difference appears to be immediately important to them. They need no theory of mind and no ability to tell where they end and the other begins to respond empathetically to the pain of the other. At the evolutionary beginnings of the view from nowhere, organisms may thus need no concept of self, no concept of other, and no capacities such as those required for recognizing themselves in a mirror.

For EMR, the capacity of capuchins to recognize and care about the pain of others would be a less well-developed form of an instinct for morality. We call this an instinct for morality because of how it is structured by a rudimentary version of the view from nowhere and because of what it enables the capuchins to detect: the form of the moral good that is tied up with caring responses to the pain of others. Moral instincts are tied to natural moral goods. Simple moral instincts allow for the detection of these goods but not for their explicit recognition. That is, the moral instinct of the capuchins does not allow them to distinguish between their goods and the moral good. For some capuchins, their own good, after a time, becomes more salient, and they pull the chain and eat. They are no doubt conflicted, and the conflict involves their own good and the good of another, but they are in all probability not aware of the conflict at this level of cognitive specificity.

Better-developed forms of the moral instinct will add this sort of complexity to the evolved capacity for morality. In later chapters, we will have more to say about these versions of the capacity; here we conclude with an immediate and obvious objection to the conflict facing the capuchins as it has already been described. Why not say that the conflict facing the capuchin in the first cage is between different goods that are both best described as its own goods? That is, there is the obvious good of quelling feelings of hunger and the less obvious good, perhaps, of quelling the feelings of discomfort that come from feeling the pain of another. This sort of argument is a familiar one to philosophers, though it is usually applied to people and not to monkeys. As such, we are not going to explore it in its full depth. But we do have some definite things to say against it, empirically and theoretically.

At the level of neurophysiology, it is an interesting question what fMRIs might or might not show. The discomfort of the first monkey might simply be discomfort over the pain of the other, as we are claiming, and not some sort of second-order discomfort over its own discomfort over the pain of the other. Any such second-order mental states are likely to be much more complicated than the first-order mental states at the level of neurophysiology, and the less complex states are likely to evolve before the combination of these states with more complicated states. This is testable, and we may be wrong. But so too for the other side, and we like our chances better than theirs on this point (see, for example, Harbaugh, Mayr, and Burghart (2007) for interesting empirical work in this area).

A related and more theoretical point is that what counts for the evolution of morality is the first-order emotional response: the monkeys are moved by the pain of another, as are the humans in the aforementioned 2007 fMRI experiment. This discomfort may make us uncomfortable in other ways as well, but what is important for morality is that we are made uncomfortable in a primary and direct way by the pain of the other. It is simply not true that we only do what we do because of some discomfort that is wholly our own. Of the many causal factors that lead to caring responses, causal factors internal to the responding organism will of course be important, from psychological discomfort to its physiological correlates. But if in addition to these proximate causal factors internal to the responding organism, there is the external causal factor of the pain and distress of another organism, acting as a causal trigger in the responding organism's environment, we are, with such a species of organism, on a moral trajectory.

On the view we are developing here, the entire causal story of certain behavioural patterns we humans recognize as connected to morality is important: as species of organisms become increasingly social and intelligent, moral trajectories take such organisms on increasingly complex moral paths, starting with moral goods such as responding to the pain of others, and then, higher on the evolutionary trajectory, instincts that enable organisms to respond in morally appropriate ways to these goods, mostly furthering them, but sometimes subverting them in the direction of moral evil. The problem with bees is that they are not on a trajectory that leads to anything we can recognize as morality. The problem with humans is that we are at a complex enough point on the trajectory that we can intentionally ignore our moral instincts. On our view, this is a naturally arising moral problem, not a problem for the nature of morality itself.

Returning to the empirical level, there is growing ethological evidence (Bekoff 2004; Allen and Bekoff 2005; Pellis, Pellis, and Reinhart 2010) to suggest that other primates and mammals are able to recognize at some cognitive level such things as other minds and fairness. There is a growing literature on animal play that suggests animals are continuously and carefully monitoring both the intentions of their playmates and the fair-making aspects of play fighting that make such interactions playful as opposed to agonistic. In addition to signalling systems that permeate play activity, there are, in canids for example, rapid exchanges of eye contact (Bekoff 2004, 505). Key to maintaining play and preventing escalation into aggression is the capacity to read the intentions of the other while at the same time making clear that one's acts are not meant to be harmful, as they would be, and would be recognized to be, outside the context of play. There is also careful attention to what sorts of aggressive actions are within or outside the bounds of fair play, and in some species attention to a 50:50 rule of aggressive actions of each playmate towards the other (Pellis, Pellis, and Reinhart 2010, 405–406). Successful play also seems to be developmentally essential for trusting and cooperative relationships among adult animals: deprived of play, adults cannot distinguish appropriate limits of social interaction with other members of their group and often respond aggressively in ways that are ultimately to their social detriment.

Admittedly, the psychological and evolutionary linkages of empathy, trust, and concern for fairness are still incompletely understood, for humans as well as in the comparative context of cognitive ethology. On the other hand, the capacities involved in the pursuit of these sorts of moral goods do seem to be connected to one another in mutually reinforcing ways, suggesting that moral goodness may be a natural kind, as we are supposing it to be. We return to this point at greater length in the next chapter.

To conclude the argument of this section, we consider an objection to the capuchin case we began with: rats behave similarly in the same sort of experimental environment. If the worry can be raised over whether capuchins are psychologically sophisticated enough to experience a genuine form of empathy, the same worry arises, with even more force, with rats. Are rats responding in an empathetic way when they resist pushing their food bar when it is directly connected to the pain of another? At this point in our understanding of cognitive neuroscience, it is hard to say. EMR has an interest in this question, because it is interested in how certain structural features of the environments of certain kinds of social species affect trait selection in these same species. To develop EMR further, this kind of question must be explicitly addressed, for rats, capuchins, humans, and other species on their own moral trajectories. At this early stage in the development of EMR, our response to the fact that rats seem to respond to the pain of others in the same way that capuchins do is that what is most interesting is *that* they do this, not *how* they do this. Rats, capuchins, and humans seem to be responding to the same structural feature of their environments in roughly the same kind of way.

There is interesting evidence to suggest that rats are not simply responding to the stress levels of other rats in perilous circumstances. Bartal et al. (2014) demonstrated that rats will free stranger rats who are trapped as long as the stranger rats are of a kind they were raised with, whether this kind is their own or another. If the stranger rat is not of a kind the free rat has been raised with, the stranger will not be freed, struggle though it might in its trapped state. For closely related work on empathy across species, including rats and humans, see Decety (2015).

Moral values as natural kinds

In our argument that morality involves a special kind of adaptation, we are assuming that morality cannot be reduced to something else. What we mean by this is that morality is not a special kind of fitness, though it contributes to fitness, and that it is not some special kind of something else, like self-interest. We are also taking morality to involve appropriate responses to peculiarly moral characteristics of our evolved environment. We are claiming that it applies not just to humans, or some other specific species, but to all species that have a particular evolutionary history, and that it is likely to evolve in animals that are social and have a reasonable degree of intelligence, so that moral situations can be recognized (and preferably thought about and systematized to some degree). This means we are talking about morality as a natural kind, probably best characterized as a naturally interacting set of properties.

A consequence of this non-reducibility of morality to something else is that moral kinds as manifested in animals will not be reducible to some other properties, especially not to particular groupings of behavioural properties, broadly subject to conditioning to shape their nature. According to EMR, moral values will lead, as features of the environments of social and intelligent organisms, to something like moral instincts. These instincts, in primates who talk and argue, will lead to the development of moral codes and moral theories. Although these codes and theories may soar well above and beyond the instincts and values that they are ultimately based upon, what makes them moral codes and theories, and what ultimately grounds their moral claims upon primates such as ourselves, are the natural moral values that lie along the trajectory of the evolutionary development of morality in the human species as a discrete and very real biological phenomenon. As members of this species, we are on a moral trajectory, perhaps nearer its beginning than we might care to think: there may be much more to morality than we are currently able to recognize, given the limits of our own evolved capacities. But again, there is no guarantee for any trajectory that it will continue on, or that the species that is on that trajectory will. Inability to fully adapt to the natural moral values in its environment could potentially lead to a species extinction (Stingl 2000).

It is important to note here that the kinds of natural moral values hypothesized by EMR are not species kinds. They are biological kinds, and so they must be contingent; but if they exist in the way that EMR supposes them to, they are not contingent in the way that biological species are. Biological species arise individually from environments that are as they are for historical reasons. These sorts of historical conditions may be repeatable, but there is nothing in the underlying structural development of the evolutionary processes in question that could be used to theoretically link such repetitions to one another. In addition to being one-off events in one-off historical circumstances, species are not stable: they emerge, mutate, and go extinct. The natural moral values at the centre of EMR, on the other hand, are hypothesized to exist as deep structural features of environments that involve social and intelligent organisms, adding to the complexities of those environments in ways that predictably create new spaces for new moral values. These spaces may not be realized in the evolutionary development of a particular species, but they will still exist as evolutionary possibilities for any species that is on a moral trajectory.

EMR is not taking moral values to be metaphysically or logically necessary, but it is supposing them to be biologically necessary. In contrast, species are both metaphysically and biologically contingent. If EMR is right, moral values arise, if they do, as biologically necessary structural features of the cooperative environments of species that are both intelligent and social. While it might be argued that no biological kinds can be natural kinds, adequately responding to this argument is tangential to our purposes here.

Moral capacities and moral values

Why suppose that the selective environments for moral capacities have included natural moral values? Street (2006) puts the point like this. We run away from predators because predators are real parts of the environments we evolve in. Predators being predators and survival being survival, organisms develop capacities for detecting and responding to predators in appropriate kinds of ways. But why suppose that moral values are real parts of the environment, like predators are? It's easy to imagine bumping into an actual predator in your environment, but how exactly do you bump into a moral value? Why not simply say the natural features of the environment are as they are – perhaps a sharp cry of pain and a quick jerk back of the shocked part of the body – and that what evolves is a capacity to pay attention to such things and to act in ways so as to prevent or minimize their occurrence? So empathetic monkeys have the capacity to respond to the pain of another monkey, but they are not thereby responding to some sort of "extra" and theoretically superfluous natural moral value in their environment.

On this more standard line of thought, a set of proto-moral and moral capacities evolves in response to particular environmental pressures, but morality isn't to be found in these underlying pressures themselves. Morality somehow resides in the capacities themselves, not in the environment.

This is an empirically tempting way to proceed, and a number of important research programs proceed in just this way. For example, Bekoff and Pierce (2009) begin their book *Wild Justice* with the announcement that they will explore three related kinds of moral behaviours, or capacities, which they label as cooperative (including altruism, reciprocity, honesty, and trust), empathetic (sympathy, compassion, grief, and consolation), and justice seeking (sharing, equity, fair play, and forgiveness). The rest of the book does discuss instances of such behaviours and some of the capacities that may motivate it, but the discussion remains for the most part anecdotal, something for which Bekoff's earlier book (2007) was criticized by Dawkins (2012, 26–27).

But there are other more empirically well-grounded efforts to understand the capacities behind such behaviour patterns, to be found, for example, in Churchland (2011), Pfaff (2007) and de Waal and Ferrari (2010, 2012). Work in this area is rapidly expanding in interesting and important empirical directions in a number of related fields, including behavioural psychology and ethology, cognitive psychology and cognitive ethology, and neuroscience. Like Bekoff and Pierce, most of this work focuses first on observed behaviour patterns and then on underlying capacities for such behaviour patterns, labelling both moral or at least proto-moral.

In Chapter 4, we will examine some of this empirical work in greater detail and further develop our argument that the evolution of the capacities in question may be better understood as caused by the general structural features of social environments that we are calling natural moral values. Here we would like to make several preliminary points about starting morality from moral capacities rather than from more general features of the environments in which these capacities evolve.

First, when it comes to capacities, how do we tell when they are of the same kinds? And why would the same kinds of capacities evolve in such a wide variety of ways across such a wide variety of species? The most immediate answer to these questions is that the capacities all evolve as similar solutions to similar environmental pressures. But then why not suppose that these pressures are what is originally patterned, as EMR does?

If the environmental pressures group themselves in certain sorts of way, why not suppose these groupings are empirically and theoretically significant? This tells us how and why moral capacities are the same kind of thing: they respond, in the right kinds of ways, to the same general kinds of negative and positive moral values as these arise in particular kinds of cooperative environments. We no longer need to worry about what is a moral capacity and what is a proto-moral capacity: the capacities start simply and grow more complex as the positive and negative moral values they respond to first appear in an environment and then become more complex. We explore this way of understanding moral values more fully in the next chapter, where we give more content to the idea of moral trajectories. On our view, current explanations of moral capacities get things the wrong way around in trying to explain moral values in terms of moral capacities or behaviours. We can make more progress in understanding morality if we suppose the explanation goes the other way around, starting with natural moral values and ending with psychological capacities.

Another significant problem with tracing morality only to capacities is that some of the capacities that motivate moral behaviour are themselves morally problematic. Consider, for example, anger and disgust. Morality draws heavily on these emotions, sometimes perhaps rightly, but sometimes, certainly, wrongly. Eating pork becomes immoral not for what might count as a good moral reason – e.g., pigs are highly intelligent animals – but because pork is pork, and pork and pork eaters are disgusting. Such people, "with the pork flesh still stuck between their teeth," are not to be trusted (Crowley 2008, 271). Worse yet is what we might believe ourselves to be morally justified in doing to pork eaters, or their equivalents in whatever group we find disgusting and are angry with, laying complete and utter waste to their families, villages, and fields.

Morality is related to what we might call tribal loyalty, or in-group loyalty versus out-group hostility. Some amount of partiality is undoubtedly morally legitimate, but how much and under what circumstances? This is an aspect of morality we will explore more fully in Chapter 7. According to some critics of evolutionary ethics, tribalism is a serious problem for any naturalistically based ethics. We argue that while tribal loyalties can be morally significant, they can also lead us down morally repugnant pathways. We argue that being clearer about both natural moral values and the human capacity that has evolved to respond to them can be part of a reasoned moral response to this sort of immoral excess.

Moral values and moral normativity

To sum up our main points so far, we think it matters both empirically and ethically where we see morality starting from: whether it is originally from moral capacities or natural moral values. Empirically, we get a more unified approach to the biological appearance of morality in the world if we suppose it to begin with natural moral values rather than moral capacities. Ethically, if we start with capacities, it is hard to see how we could ever get from "is" to "ought" in any naturalistically based morality. The mere fact that we possess certain kinds of evaluative psychological attitudes that developed as they did because of their survival value in environments we found ourselves in does not give us any way of justifying the content of those attitudes. We might think that it is good to respond compassionately to the pain of others, but that does not make it morally good to respond to the pain of others. Being inclined to think that something is so does not make it so. And if it is not even *morally good* to respond compassionately to the pain of others, it is hard to see how it might be the case that we *ought* to act in such a way or be *obligated* to act in such a way by something that might justifiably be called *an overriding moral reason*.

If this is where evolutionary naturalism leaves us when it comes to morality, evolutionary naturalism might not seem all that ethically interesting. Better perhaps to look for morality elsewhere, in Kantian rationality or in some sort of social contract theory, for example. Where other forms of evolutionary ethics would disconnect moral explanation and moral justification, EMR suggests a naturalistic way in which they might be tightly joined together.

Moral truth is ultimately to be found in the natural moral values that EMR takes to be deep structural features of certain kinds of environments involving social and intelligent creatures. Moral normativity, in its most fully developed form that we know of, is to be found in the explicit moral norms that are the separate bits and pieces of the moral forms of reflective equilibrium that culturally develop among different groups of humans as they seek over time to justify their moral judgments. The key idea linking moral truth to explicit moral norms is to be found in the thought that the natural moral values that emerge at the level of the evolutionary trajectories of species that are both social and intelligent may turn out to be deeply implicated in some of the more significant human moral arguments and agreements as they emerge in processes of reflective equilibrium. A closely related idea is that moral values, and moral normativity, may be much more general biological phenomena than human moral values and human moral norms might seem to suggest. Not all moral norms need be supposed to be explicit moral norms, or perhaps better, not all forms of moral normativity need be supposed to be grounded in explicit moral norms.

An important antecedent of EMR as we are developing it here is John Dewey's pragmatic theory of values as goals of actions. According to John Dewey's criticism of the reflex arc concept in the mechanistic psychology of the late nineteenth century, organisms are not simply being impacted by stimuli that happen to produce some sort of reflexive response in and from them. They are instead actively seeking stimuli that are connected to things in their environment that it is important for them to interact with (Dewey 1896; Bredo 1998; Barrett 2011, 94–101). For example, in pursuit of nutritional goods, organisms are actively looking for things to eat. Organisms move around their environment, seeking stimuli that might be connected to edible things in that environment. When they detect food and eat it, an organic circuit is completed. The mechanistically understood "reflex arc" between the stimulus and the response is merely a disjointed representation of a larger organic circuit linking the organism to its environment.

The reflex arc more broadly understood as completing an organic circuit introduces for Dewey a naturalized concept of normativity. In an organic circuit, the right kind of object is naturally linked to the right kind of biological imperative: "eat this!" in the example we are considering related to nutritional goods. EMR hypothesizes that moral goods are linked to natural moral imperatives of a similar kind: help someone in need of help, play fair, and so on. What makes these sorts of biological imperatives moral imperatives is the kind of good thing in the environment that they are linked to. What makes them imperatives is the evolution of an organic circuit that links the right kinds of things in an organism's environment to the right kinds of actions, such as actions the organism ought to perform in pursuit of what is good for it.

This account needs to be complicated a bit to accommodate morality as instinctual, but the change is fairly minor, at least conceptually. Rather than the completion of a circuit, there is a satisfaction of an instinct. This strengthens the connection of the instinct with the kind of behaviour involved, where kinds are not necessarily merely generalizations of actions but are determined by the patterned character of the instinct, which is itself patterned by the value it arises in response to. For the functionality of the instinct to be preserved, organisms "ought" to do what fits the reinforced pattern. The normativity of natural values is built into the evolutionary context in which these values arise. As a key hypothesis, this is where EMR places the origin of moral normativity. EMR does not deduce moral normativity from what is natural but instead supposes that moral normativity is originally part of the natural world.

Nutrition and predation are particularly deep structural aspects of the biological world. We might thus suppose, following Dewey, that an important part of the design build of organisms will be mechanisms for searching out both kinds of things, possible nutritional goods in the environment as well as possible bad predatory things. These mechanisms, to successfully develop, will need to be closely tied to responses that are appropriate to the kind of good or bad thing in question.

What does this sort of biological normativity fundamentally have to do with what we might philosophically think of as "the real thing," namely human moral normativity? De Sousa (2017) argues that even if nature had its own purposes for us, moral or otherwise, this would not in and of itself tell us what our own individual or social purposes ought to be. On any naturalistic approach to morality, this much will certainly be true. The point of EMR, however, is that moral values are a deeply important part of our environment as the highly developed social and intelligent primates that we are. Whatever our own purposes, moral values are deep structural aspects of our environment we are likely to be interested in, perhaps to use them against others and ultimately against themselves as we attempt to climb to positions of social power, or alternatively, to resist such efforts to consolidate social power in these oppressive and hegemonic sorts of ways. We may of course try to keep ourselves free of such messy social affairs and, as part of this, try to ignore moral values as much as we are able to. This is likely to limit us considerably and would most likely leave us as easy prey for those who would like to make use of us for their own purposes. From a biological point of view, free will is both a blessing and a burden. Either way, it does leave humans with some amount of room to do as we please.

To take the question of human purposes a step further, we might consider a later branch of pragmatism as represented by the work of Richard Rorty. Rorty (1987) makes a general argument that there is a single concept of truth that applies to the humanities, the social sciences, and the so-called hard sciences. Truth is where non-coercive and fully inclusive reasoned discourse leads us when we search for consistency and coherence among our many beliefs, whether we are doing physics or philosophy. Wherever the search for greater consistency and coherence might take us, where no views are privileged and none are ignored, that is where the truth lies. There is no external reality beyond the reach of our uncoerced and thus impartial ability to articulate and justify through reasoned discourse theories and statements about the way things are. Scientific truth is thus in principle no more objective than philosophical truth. Neither form of truth tracks an external world of facts that exists independently of reasoned human discourse.

Daniel Dennett (2007, 256) provides an alternatively pragmatic and precisely targeted reply to this position. All biological organisms are playing hide and seek in a dangerous world of other organisms. Human science and technology have elevated us to a safe enough position that we tend to forget about this fundamental vulnerability that we share with all other organisms. But there is a real biological world of which we are a part, and it is not overly friendly. If we overuse antibiotics microbes will come back to eat us, whether we are aware of this or not. The scientific method requires us to reach out into the world itself to try and touch what exists there by making predictions that can be experimentally tested. The observational statements that result from experiments need to be made consistent with other statements, both observational and theoretical, but they are generated by theories telling us to look for things in places and ways we otherwise would not have thought to have looked. In science, it matters whether our consistent and coherent theories are on the track of the truth about the world we inhabit, a truth that exists independently of and externally to our theoretical arguments, predictions, and statements about its fundamental nature. The germ theory of disease, for example, told us to look for experimental results in places we would otherwise not have looked, and we need to continue to pay attention to this theory and its predictions, at least until a better theory with better predictions comes along.

According to EMR, moral goods may be just as important as predators in a social and intelligent creature's natural environment: ignore them and you are likely to perish. For some creatures, playing fair, helping others and cooperating in joint tasks towards common goods may be vital to their survival. Moral goods are thus very real aspects of the natural biological world. Not all organisms may be able to detect them, and not all organisms may need to be able to detect them. But if they are in your environment and they matter to your well-being, you are likely better off, in terms of your own overall good, if you are able to detect them and respond appropriately.

Note

1 An earlier version of this chapter appeared as "Evolutionary Moral Realism," *Biological Theory* 7 (2013): 218–226. We would like to thank Springer Nature for permission to reprint this material here.

Bibliography

Alexander, Richard D. 1987. *The Biology of Moral Systems*. New York: Aldine de Gruyter.

Allen, Colin, and Marc Bekoff. 2005. "Animal Play and the Evolution of Morality: An Ethological Approach." *Topoi* 24 (2):125–135.

Atran, Scott, and A. Norenzayan. 2004. "Religion's Evolutionary Landscape: Counterintuition, Commitment, Compassion, Communion." *Behavioral and Brain Sciences* 27:713–770.

Barrett, Louise. 2011. *Beyond the Brain: How Body and Environment Shape Animal and Human Minds*. Princeton: Princeton University Press.

Bartal, Inbal Ben-Ami, David A. Rogers, Maria Sol Bernardez Sarria, Jean Decety, and Peggy Mason. 2014. "Pro-Social Behaviour in Rats Is Modulated by Social Experience." *eLife*. http://dx.doi.org/10.7554/eLife.01385.001.

Bekoff, Marc. 2004. "Wild Justice and Fair Play." *Biology and Philosophy* 19:489–520.

Bekoff, Marc. 2007. *The Emotional Lives of Animals: A Leading Scientist Explores Animal Joy, Sorrow and Empathy – and Why They Matter*. Novato, CA: New World Library.

Bekoff, Marc, and Jessica Pierce. 2009. *Wild Justice: The Moral Lives of Animals*. Chicago: University of Chicago Press.

Bredo, Eric. 1998. "Evolution, Psychology and John Dewey's Critique of the Reflex Arc Concept." *The Elementary School Journal* 98 (5):447–466.

Brosnan, S.F., and Frans B.M. de Waal. 2003. "Monkeys Reject Unequal Pay." *Nature* 25 (18):297–299.

Churchland, Patricia. 2011. *Braintrust: What Neuroscience Tells Us about Morality*. Princeton: Princeton University Press.

Crowley, Roger. 2008. *Empires of the Sea: The Final Battle for the Mediterranean*. London: Faber and Faber.

Darwin, Charles. 1874. *The Descent of Man and Selection in Relation to Race*. 2nd ed. London: John Murray.

Dawkins, Marian Stamp. 2012. *Why Animals Matter: Animal Consciousness, Animal Welfare, and Human Well-Being*. Oxford: Oxford University Press.

Dawkins, Richard. 2006. *The Selfish Gene (30 Anniversary Edition)*. New York: Oxford University Press.

Decety, Jean, Inbal Ben-Ami Bartal, Florina Uzefovsky, and Ariel Knafo-Noam. 2015. "Empathy as a Driver of Prosocial Behaviour: Highly Conserved Neurobehavioural

Mechanisms across Species." *Philosophical Transactions of the Royal Society B* 371:1–11. http://dx.doi.org/10.1098/rstb.2015.0077.

Dennett, Daniel C. 2007. *Breaking the Spell: Religion as a Natural Phenomenon*. New York: Penguin.

de Sousa, Ronald. 2017. "Nature's Purposes and Mine." In *How Biology Shapes Philosophy: New Foundations for Naturalism*, edited by David Livingstone Smith, 141–160. Cambridge: Cambridge University Press.

de Waal, Frans B.M. 1989. *Peacemaking among Primates*. Cambridge: Harvard University Press.

de Waal, Frans B.M., and Pier Francesco Ferrari. 2010. "Toward a Bottom-Up Perspective on Animal and Human Cognition." *Trends in Cognitive Sciences* 14:201–207.

de Waal, Frans B.M., and Pier Francesco Ferrari, eds. 2012. *The Primate Mind: Built to Connect with Other Minds*. Cambridge, MA: Harvard University Press.

Dewey, John. 1896. "The Reflex Arc Concept in Psychology." *The Psychological Review* 3 (4):357–370.

Essler, Jennifer L., Sarah Marshall-Pescini, and Friederike Range. 2017. "Domestication Does Not Explain the Presence of Inequity Aversion in Dogs." *Current Biology* 27 (12):1861–1865.

Fletcher, G.E. 2008. "Attending to the Outcome of Others: Disadvantageous Inequity Aversion in Male Capuchin Monkeys (Cebus Apella)." *American Journal of Primatology* 70 (9):901–905.

Gould, Stephen Jay. 1977. "So Cleverly Kind an Animal." In *Ever Since Darwin: Reflections on Natural History*, 260–267. New York: W.W. Norton & Company.

Harbaugh, W.T., U. Mayr, and D.R. Burghart. 2007. "Neural Responses to Taxation and Voluntary Giving Reveal Motives for Charitable Donations." *Science* 316: 1622–1625.

Hauser, Marc D. 2006. *Moral Minds: How Nature Designed Our Universal Sense of Right and Wrong*. New York: HarperCollins.

Herberholz, J., and B. Schmitz. 1998. "Role of Mechanosensory Stimuli in Intraspecific Agonistic Encounters of the Snapping Shrimp (Alpheus Heterochaelis)." *Biological Bulletin* 195:156–167.

Holldobler, B., and E.O. Wilson. 2008. *The Superorganism: The Beauty, Elegance, and Strangeness of Insect Societies*. New York: W.W. Norton.

Hughes, M. 1996. "Size Assessment Via a Visual Signal in Snapping Shrimp." *Behavioral Ecology and Sociobiology* 38:51–57.

Joyce, Richard. 2001. *The Myth of Morality*. Cambridge: Cambridge University Press.

Mackie, John Leslie. 1977. *Ethics: Inventing Right and Wrong*. Harmondsworth: Penguin.

Nagel, Thomas. 1986. *The View from Nowhere*. Oxford: Oxford University Press.

Nichols, Shaun. 2004. *Sentimental Rules: On the Natural Foundations of Moral Judgment*. Oxford: Oxford University Press.

Pellis, Sergio M., Vivien C. Pellis, and C.J. Reinhart. 2010. "The Evolution of Social Play." In *Formative Experiences: The Interaction of Caregiving, Culture and Developmental Psychobiology*, edited by C. Worthman, P. Plotsky, D. Schechter and C. Cummings. Cambridge: Cambridge University Press.

Pfaff, Donald D. 2007. *The Neuroscience of Fair Play: Why We (Usually) Follow the Golden Rule*. New York: Dana Press.

Rachlin, Howard. 2000. *The Science of Self-Control*. Cambridge, MA: Harvard University Press.

Range, Friederike, Lisa Horn, Zsofia Viranyi, and Ludwig Huber. 2009. "The Absence of Reward Induces Inequity Aversion in Dogs." *PNAS* 106 (1):340–345.

Rorty, Richard. 1987. "Science and Solidarity." In *Rhetoric of the Human Sciences: Language and Arguments in Scholarship and Public Affairs*, edited by John S. Nelson, Allan Megill and Donald N. McCloskey, 38–52. Madison: University of Wisconsin Press.

Ruse, Michael. 1986. *Taking Darwin Seriously: A Naturalistic Approach to Philosophy.* Oxford: Blackwell.

Singer, Tania, Ben Seymour, John O'Doherty, Holger Kaube, Raymond J. Dolan, and Chris D. Frith. 2004. "Empathy for Pain Involves Affective But Not Sensory Components of Pain." *Science* 303:1157–1162.

Stingl, Michael. 1996. "Evolutionary Ethics and Moral Theory." *The Journal of Value Inquiry* 30:531–545.

Stingl, Michael. 2000. "All the Monkeys Aren't in the Zoo: Evolutionary Ethics and the Possibility of Moral Knowledge." In *Moral Epistemology Naturalized*, edited by Richmond Campbell and Bruce Hunter, 245–265. *Canadian Journal of Philosophy, Supplementary*, Vol. 26.

Stingl, Michael, and John Collier. 2004. "After the Fall: Religious Capacities and the Error Theory of Morality." *Behavioral and Brain Sciences* 27 (6):751–752.

Stingl, Michael, and John Collier. 2005. "Reasonable Partiality from a Biological Point of View." *Ethical Theory and Moral Practice* 8:11–24.

Street, Sharon. 2006. "A Darwinian Dilemma for Realist Theories of Value." *Philosophical Studies* 127 (1):109–166.

van Wolkenten, M., S.F. Brosnan, and Frans B.M. de Waal. 2007. "Inequity Responses of Monkeys Modified by Effort." *Proceedings of the Natural Academy of Sciences* 104 (47):18854–18859.

Wilson, Edward O. 1998. "The Biological Basis of Morality." *Atlantic Monthly* 281 (4):53–78.

3 Moral trajectories

Moral goods as real things

A key part of what takes EMR from pistol-toting shrimp to pistol and ought-toting humans is what we are calling moral trajectories. In this chapter, we give more empirical content to this idea. This will lay groundwork for the second half of the book, where we face two significant philosophical problems: whether EMR commits some version of the is/ought fallacy, and how exactly EMR is supposing natural moral values may be related to moral reasons and moral responsibility.

As we noted at the end of the last chapter, some of the ideas we develop in this book are similar to John Dewey's naturalistic approach to ethics in the early twentieth century. Dewey's ideas in this area are more suggestive than fully developed, and as we ourselves endeavour to develop the central ideas of EMR, we do not see an immediate need to discuss Dewey's work much more than we already have. We mention Dewey at this point in our argument to contrast the sort of evolutionary approach to ethics originally suggested in Dewey and Tufts' (1908) *Ethics* with G.E. Moore's (1903) *Principia Ethica* and the long epistemological and metaphysical shadow this latter book cast over twentieth-century ethics in the Western philosophical and scientific world (Stingl 1997).

Moore began twentieth-century ethics, in the Anglo-American philosophical world, with the bold and influential claim that moral values were real but non-natural objects, something like mathematical objects or Platonic forms. If moral language was meaningful, such objects had to exist, to give meaning to moral terms in the only way they could get their meaning. Moore arrived at this conclusion using the open question argument. For any definition of a moral value like goodness, natural or otherwise, we could always raise the question, in a meaningful way, of whether the thing or things that met the given definition were in fact morally good. Mill, Moore thought, had naturalistically defined the good as the pleasurable, and Moore thought we could then sensibly ask whether pleasure was, in fact, morally good. Given the open nature of this question, whatever we might mean by "moral goodness," it would have to be something other than "that which is pleasurable," with the same sort of argument applying to whatever other kind of thing we might try to define moral goodness in terms of, whether it be natural or non-natural. Thus, Moore thought that "moral goodness" could not be defined, and hence the thing that it referred to, moral goodness, must be a simple, non-natural property of whatever things the term rightly applied to.

For early twentieth-century ethics, this conclusion, based on what appeared to be a powerful theory of language and meaning, created two pressing questions: how might knowledge of such values be possible, and how could knowledge of such values be connected to human reasoning, motivation, and behaviour?

Moore's own answer to the epistemological question was an appeal to moral intuition, an appeal further developed by W.D. Ross and other early intuitionist philosophers. But the appeal to moral intuitions remained explanatorily opaque, and an obvious answer to its opacity appeared in the moral scepticism of the logical positivists. For the positivists, and the expressivists who followed, what gave moral claims their meaning was not a descriptive aspect of language but an expressive aspect. When we said that things were morally good or bad, we weren't making a descriptive claim about the real world, non-natural or otherwise, but were expressing a positive or negative attitude about the thing we were calling good or bad. The intuitionistic approach to moral goodness, "Whatever it is, I know it when I see it," thus became the much more tractable "I know what I like."

In addition to dissolving the epistemological problem posed by real but non-natural moral values, this positivistic move in the philosophy of language simultaneously solved the motivational problem that faced Moore and the intuitionists. Attitudes, unlike propositional thoughts, are exactly the sort of mental state that motivates creatures like humans to act. What we like is typically what we pursue. If values were real objects, then even if they were more epistemologically accessible objects, there would remain the very real question of how these objects get their motivational power and authority over us. As John Mackie (1977, 38–42) put the point in his so-called Argument from Queerness, it is one thing to suppose that we will ever have knowledge of anything like the Platonic forms, and quite another thing, even more bizarre, to suppose that once we were to have such knowledge, it would somehow in and of itself motivate us to act. Although we will not pursue the matter here, we think the attraction to arguments like Harman's and Street's is a continuing part of the philosophical legacy of logical positivism as an empirically plausible response to the moribund project of Moore's moral realism.

Moral goods and the interests of others

In this chapter, we wish to remain less philosophically encumbered and to develop at a more empirical level the basic idea of what we regard as a plausible evolutionary theory of morality, one that starts from the simple sorts of moral values mentioned in earlier chapters. Morality is not merely the province of philosophers. Moral values are unlikely to be some sort of non-natural object, accessible only to human intelligence and reasoning. Moral values are similarly unlikely to have only come into existence through the expressive powers of human language: they are too deeply embedded in the forms of cooperation that led to the evolution of human intelligence and language.

Central to our biological approach to morality is the idea of moral trajectories. To better understand this idea, we want to return to the question of how such trajectories might get started. To pursue this question further, let us return to self-interest, a natural phenomenon that philosophers have not been so metaphysically or epistemologically puzzled by. For humans, we think that certain things are, as a matter of fact, in our self-interest, that it is possible to have knowledge about such things, and that such knowledge is far from motivationally inert. I have an interest, call it **I**, I come to know of **I**, and this knowledge leads me to act to satisfy **I**.

We can quibble, philosophically, about whether interests require language, but in more general terms, we have little difficulty in imaging that other organisms have their own interests, the interests that they spend most of their lives pursuing. How other organisms are actually attracted to the things that are in their interests and how these attractions trigger appropriate behaviours are interesting and important questions of behavioural psychology, but they aren't matters of deep metaphysical doubt.

For certain forms of cooperation, EMR supposes that the interests of others are also attractive, some of the time. That I eat a worm is in my immediate nutritional interest. That I put it in your mouth, and you eat it, is immediately in your nutritional interest. This is how morality enters the biological world, with organisms that take an active and positive interest in the interests of others.

Or perhaps this way of talking is overblown: simple organisms are hardly *taking an interest* in anything, never mind an interest in an interest of another. However, the point is that some things are directly in their interest, in their environment, and that some of the things that are in the direct interest of a related organism are also things that they cue to, with some mechanism or other, in such a way that they respond to them with behaviours that bring about the satisfaction of the direct interests of others.

Some will argue that interest is everywhere and always based on self-interest: if I take an interest in others, it is this interest that must be metaphysically and motivationally prior, because, after all, it is mine. But again, why suppose this more complicated, two-tiered story is likely to be true? Why not suppose that in the beginning there are simply interests in an organism's environment, some more immediately those of the organism and some more immediately those of other organisms? Even calling such attractors interests is already stretching things, because in the case of something like male pistol shrimp, defending territory means, if it means anything, responding to a particular trigger in their environment in a particular way. What is immediately important for our argument is that they might be primed to be receptive to such environmental triggers and to behave in certain ways relative to those triggers. Perhaps such receptivity and behaviours further what might be described as their self-interest, but perhaps in ways that might be described as also furthering interests in group cooperation.

We might also think that all interests are self-interests within a one-tiered, not two-tiered, approach. That is, instead of arguing that I have a primary self-interest and a derivative interest in others, we could hold that I simply have two interests. I could put the worm in my mouth or put the worm in your mouth, and if the right mouth is positioned in the right way, that's the mouth I put the worm in. This may be a fine way of talking, but morality, as we are approaching it here, has snuck back into the picture. I could have put the worm in my own mouth, but I didn't: I put it into yours. Not because I thought that was better for me, but because your mouth was more salient or otherwise more important than mine in the environmental context I was in. In this context, it was better for the worm to go into your mouth than into mine. Not just better for you or for me, just better. Morally better.

Moral values are related to cooperative and reproductive success but are not the same thing as cooperative and reproductive success. We say "morally better" because the goodness involved is of a particular kind, a kind we have come to recognize, with our much more cognitively well-developed moral capacities, for what it is.

Some basic moral goods

Here are some empirical examples of what we mean by saying that moral values are the kind of thing that individual organisms may become aware of in their surroundings. One important way of responding to the interests of another is by helping that other individual reach an end that he or she cannot reach on his or her own. Helping others who need your help seems, intuitively, to be a fairly basic moral good. It also seems to be the kind of thing that both chimps and human infants as young as 14 months old can be aware of and responsive to.

In a series of experiments (Warneken and Tomasello 2006; Warneken et al. 2007; Warneken and Tomasello 2007), chimps and human infants were exposed to a situation in which someone was trying to reach for something that he or she had accidently dropped, something that was a necessary part of an activity he or she had been engaged in, such as a clothes pin (the experimenter was hanging things on a line) or a marker (the experimenter was drawing pictures). In a control situation, the experimenter purposely threw down the object; in the experimental situation, it was accidently dropped. Both the chimps and the infants noticed the accidently dropped object and the reaching, and both responded by returning the object to the experimenter. They did this whether or not they were rewarded, and in circumstances where reciprocation was ruled out by the circumstances of the experiment. They also did more of it as they were exposed to more experimental situations of a similar kind and presumably understood better what the experimenter's problem was.

We might say here that the chimps and infants had an interest in helping the experimenter, but they themselves seem not to have had this interest until they discovered it. And whatever they discovered, it doesn't seem to be an interest of theirs as much as a salient feature of their environment, namely that someone needed help to reach an end he or she was engaged in trying to reach. This feature of their environment emerged as salient to them in the course of the experiment, and it was something they both noticed and responded to.

As we noted in Chapter 2, Dewey calls this sort of connection between a significant feature of an organism's environment and the organism's predisposition to respond to it appropriately an organic circuit. The fundamental idea of the organic circuit hypothesis is that animals (with or without behavioural flexibility) are looking for particular kinds of things in their environment to do things with, most immediately things like predators, prey, reproductive partners, or off-spring that need to be fed. The biological circuit may be instinctual, learned, or a combination of the two. What makes particular features of the environment interesting and important is that they require doing something with, and the animal is primed, by the circuit, to do whatever is called for by the kind of thing in question, when it encounters this thing. In the organic circuit, the motivation to act is built into the initial recognition of the thing that triggers the animal's response; moreover, the animal notices certain kinds of things precisely because they call for important kinds of actions in the animal's environment. The organism involved may have some level of flexibility in whether it will respond appropriately or it may not. The circuit may be tighter or looser, depending on the degree of sophistication in the organism's motivational system.

Helping someone reach an end he or she is striving for responds directly to the interests of another. Let us call helping someone in this sort of situation a first-order moral good. The chimps and infants we have been talking about become aware that someone needs help and they help. To this extent, they are aware of this kind of moral good in their environment. But there is additional evidence that their awareness of this good is more sophisticated than this first set of experiments might suggest, and that awareness of first-order moral goods is linked to awareness of higher orders of moral goodness.

In another series of experiments, infants as young as three months old were exposed to puppets or animated figures who either helped or hindered another puppet or animated figure reach an end it was trying to reach.

In Hamlin, Wynn, and Bloom (2007), a wooden figure with eyes glued on it was trying to get up a hill while 6- and 10-month-old infants watched. A helper figure either helped it get up, or a hindering figure pushed it back down. The 10-month-old infants watched the hinderer for longer periods of time, but both 6-month-olds and 10-month-olds preferred the helping figure when offered a choice between it and the hinderer. Hamlin et al. noted that the infants had not interacted with any of the figures previously, and that the infants seemed to be reacting to the social aspects of the situation they were watching rather than superficial perceptual aspects of the situation. In a control arm of the experiment, inanimate objects (no eyes or independent movement) were pushed up or down the hill, following the same trajectories as in the earlier arm of the experiment. In this case, infants showed no preference for either of the figures who pushed the object either up or down the hill. In a third arm of the study, involving a neutral figure that neither helped nor hindered the figure trying to get up the hill, infants preferred the helper figure to the neutral figure and the neutral figure to the hindering figure when offered two-way choices of figures.

In Hamlin and Wynn (2011), five- to nine-month-old infants were shown puppets who either helped or hindered a third puppet open a box with a brightly coloured rattle inside. In an echo of the Warneken studies discussed above,

Hamelin and Wynn also showed five- to three-month-old infants a situation where a puppet dropped a ball and a giver or taker puppet either gave it back or kept it. As in the earlier experiments, the infants preferred the helping puppets to the hindering puppets when offered a choice between them in both of the additional sets of experiments.

Connecting concern with fairness to concern with helping others, Anderson, Takimoto et al. (2013) had capuchins watch two human experimenters trading balls back and forth. In the first experimental situation, each experimenter started with three balls in a clear container along with a second clear but empty container. The balls and the experimenters were new to the capuchins. To begin the experiment, one experimenter would hold out his or her empty container to the other, who would fill it with his or her three balls. When that second experimenter held up his or her own empty container to the other, the other would either transfer balls or not. The two experimenters would then offer food to the capuchins, who strongly preferred taking it from reciprocating experimenters than from non-reciprocating experimenters. In other experimental conditions, where only one of three balls was transferred in response to three balls having been initially transferred (incomplete reciprocation), or where only one ball was transferred in response to an initial transfer of three balls (impoverished starting points), the monkeys preferred taking food from the experimenter who was impoverished and only transferred one ball compared to the experimenter who had three balls but stopped transferring after only one of them. This suggests that the capuchins valued helping behaviour, but paid attention to the position of the helper. This suggestion was further tested in Anderson, Kuroshima et al. (2013), in which capuchins watched pairs of individuals where one was in a position to help the other or was alternatively preoccupied in a task of their own. The monkeys did not discriminate against non-helpers when they were otherwise occupied.

If we assume that helping other individuals when they need it is a moral good, infants and other primates seem capable of noticing at least three important things regarding this moral good.

First, they can recognize it from the inside: when someone needs help, they notice it and they respond. Second, they can also recognize this good from the outside: when someone needs help, they prefer helping behaviour to hindering behaviour when a third party is in a position to do the one or the other. And third, they also respond to the helper or hinderer: the hinderer did not do what it ought to have done, and now it is something itself to be avoided. Not doing the morally good thing is morally bad.

Evolutionary development of moral values and moral normativity

What we are supposing is that directly responding to the non-moral good of another is a basic kind of moral good. Part of what makes this approach to moral goodness seem implausible is the fact that it seems to extend morality as far back, perhaps, as pistol shrimp and to organisms that put worms into others' mouths because these other mouths are in some contexts a more salient aspect of their environment than their own. The apparent implausibility of our approach is partly, we think, due to an anthropocentric assumption that morality requires responsibility and more deeply, a conceptually sophisticated form of normativity. It is also, we think, due to the deep moral scepticism of the twentieth century and the alternative forms of moral realism or moral objectivity that take moral goods to be some sort of non-natural kind of thing or human creation. And it is, in no small part, due to worries about the naturalistic fallacy, worries which are directly connected to issues of normativity.

In this section, we want to begin to explore, still in a philosophically naive sort of way, what moral normativity might amount to on our evolutionary approach to morality. Instead of beginning with metaphysical worries about the possible existence of natural moral goods, the beginning question for us is an empirical one: does assuming the existence of such goods have explanatory value? If so, they are plausible, and whatever metaphysical arguments might be marshalled against them are of diminished interest.

A key point of this section is that the right, that which ought to be done, arises directly from the good, the dominant (positive) moral value in a given environmental context. We mean this as a general empirical and moral truth, one which needs much refinement and careful qualification. But it is, we think, a basic truth about morality, however crudely we are stating it here.

We think that the existence of natural moral goods is a plausible empirical hypothesis, which we take to be the principal hypothesis of EMR. The question is how such goods might give rise to moral normativity. Our basic answer is two-fold. First, basic moral values give rise to more complex forms of moral value. Second, as moral values become more complex, this complexity brings with it increasingly complex forms of normative force. By normative force, we mean pressures from within the environment that push the organism to act in certain sorts of ways. As organisms develop more complex patterns of moral behaviour, more of these pressures will arise and more of them will become more deeply internalized within the organism's own motivational system. We call this particular kind of normativity moral because it starts with basic moral values and culminates, as far as we know from where we are on our own moral trajectory, with the articulation and implementation of moral norms.

Along our own trajectory, a late-arriving moral value is that of autonomy. We are not always as autonomous as we think we are: those who walk past a bakery with the smell of fresh pastry in the air are more likely to offer spontaneous help to someone who has dropped a glove than others (Gueguen 2012). Nevertheless, we are relatively self-directing, especially in comparison to other animals. As a moral value, autonomy, like all later moral values, is linked to earlier moral values. Take responsibility. Autonomy and responsibility, as moral capacities, emerge on our trajectory in moral evolution. But they develop from, and in mutually supportive relationships with, other moral values, values which guide organisms in moral directions, even if this guidance is not as fully autonomous as it is at our own level. In general, we see moral capacities, such as moral emotions, as the internalization of the moral force of moral values as they come to exist in the biological world. Wherever organisms are guided by moral values, we think there is normative governance of those organisms by the values in question. For us, this is the beginning of the moral force of moral values on organisms and thus the beginning of moral normativity. Moral normativity may culminate, in our case, with the ability to articulate, implement, and consciously act upon moral norms, but it is not reducible to this particular endpoint.

Wherever there is behavioural flexibility and purposeful behaviour, there will be pressures exerted on organisms to behave in some ways rather than others. When these pressures involve putting the interests of another before one's own, we have the beginnings of moral normativity. But even where there is no flexibility for individual organisms, a beginning level of normativity may be found in the organic circuit itself. The circuit is selected for because it links the detection of something in the environment to a particular kind of action: towards the thing in the environment if it is a helpful kind of thing and away from it if it is a harmful kind of thing. Where the kind of thing in question is moral in nature, we have the beginnings of moral normativity.

In the preceding section, we focused on a basic kind of moral value, helping someone who needs to be helped. Here we want to explore more complex ways of responding to the interests and needs of others as well as navigating the shared space between those interests and needs and one's own needs and interests. More complex moral values arise with more complex organisms, but they arise from more basic moral values and are supportive of those more basic values. With the advent of more complex organisms comes more behavioural flexibility and with more flexibility more pressures, both external and internal, to act as one morally ought to.

Our first example was briefly mentioned in Chapter 1: fair play in the context of rough-and-tumble juvenile play fighting. Fairness is what appears to be the distinguishing mark between play and aggression (Pellis and Pellis 2009, 42–45). Rats and other rodents, for example, appear to follow a rough sort of 50:50 rule in their juvenile play fighting. Attackers will make an attacking move but without the defensive, protective move that would accompany it in a real fight, and in response to such an attack, the defender will move more slowly in responding to the unguarded attacking move with its own defending move. An attacking juvenile rat, for example, instead of pinning an opponent it has gotten down will partially pin it down, using an unstable stance that makes counter-attack easier. No one is playing for keeps, and as much as one presses one's own attack, one must leave oneself open to counter-attack.

Apparent breaches of fairness may signal the end of play and the onset of genuine aggression. Or not, and animals need ways of telling which is which and what is what (Pellis and Pellis 1996). In degus, for example, a playful sequence of attacks and counter-attacks that threatens to get out of hand can remain playful if the overly aggressive attacker offers himself or herself as a target for return (and playful) aggression from the recipient of the vigorous play attack that might otherwise threaten to end the playful exchange (Pellis, Pellis, and Reinhart 2010,

411). Degus can deliver powerful kicks to one another with their hind legs, and an animal that has been successfully kicked tumbles over and becomes an easy target for a quick bite. In delivering such a kick, even in play, the aggressive animal threatens to go beyond the bounds of fair play, potentially appearing to do something it ought not to have done in the context of play fighting. For the encounter to remain playful, the animal that delivered the kick must now do something to re-establish the importance of those boundaries to the continued interaction with its playmate, if, of course, the interaction really was a playful one.

It must make some sort of gesture of appeasement, a signal of "just playing," a signal it makes in this case by not pressing its attack further, with a bite, as it would in a real fight and even going so far as to offer parts of its body as easy targets for a playful bite back. And for the fight to remain playful, the other animal must respond to, or accept, this gesture of appeasement. It would go too far here to suggest that the first animal is asking for reconciliation or forgiveness and the second animal bestowing it. But with appeasement – or something prior even to that, if that is what we are dealing with in this case – we have another fairly basic kind of moral value, this time tied to boundaries that both animals have to stay within and to some degree recognize. The boundary transgressed here is a moral boundary, and hence the normative force of the initial boundary and the restoration its transgression calls for is morally normative.

Is this "appeasement"? Something was noticed that was unfair in the play environment, and fixing it required a noticeable moral good. In more sophisticated animals, who also play fairly, appeasement and reconciliation appear in this same kind of context. In even more sophisticated animals, like humans, forgiveness appears, again in the same kind of context. Someone went too far into the zone of someone else's interests, and something has to be done to set matters right. Reconciliation and forgiveness appear to be more complex forms of the original kind of moral good that appears at least as early as rats. As this kind of moral good gets more complex, as in humans, the pressures mount to do what one ought to. Initiating reconciliation requires gestures of appeasement, and failing to make such gestures threatens the social cohesion of the group – and everyone is uneasy in the absence of such a gesture. The pressures to forgive are more complicated still, but they are the same kinds of pressures in the same kinds of contexts. To forgive may well require norms regulating blame and guilt, but on our view, these norms are getting their initial moral push from outside the process of articulating and implementing them.

Not all aggression is playful, of course. But in social mammals, much of it occurs in the context of ongoing relationships within groups of kin and non-kin. In primates, there is a large and growing literature on reconciliation after in-group aggressive encounters (Aureli and de Waal 2000). Reconciliation has also been studied in goats, hyenas, and dolphins. In primates, depending on the species, it includes mouth kissing, embracing, sexual intercourse, clasping the hips of the other individual, grooming, grunting, and extending and holding hands (de Waal 2004). In some species, such as chimpanzees, it may involve a third individual who mediates between the two individuals who just experienced an aggressive encounter with one another, and it may involve the entire group, as in an incident in the Arnhem Zoo in which the alpha male violently attacked an older female. After other group members broke up the encounter with screaming and chasing, there was a tense silence that was broken by group hooting. In the pandemonium that followed, the two antagonists kissed and embraced (de Waal 2000, 586).

Dogs are much less cognitively sophisticated than chimpanzees, but there is evidence that dogs reconcile and that this reconciliation can sometimes involve third parties (Cools, Van Hout, and Nelissen 2008). Third-party affiliative behaviour towards victims of an aggressive encounter has also been observed in rooks, as has affiliative behaviour directed towards former opponents (Seed, Clayton, and Emory 2007). It is unclear to what degree, if any, rooks or dogs are capable of genuine empathy, which seems to be part of affiliative and consoling behaviour in primates. Dogs or rooks may thus be interesting intermediate cases of the ability to be able to detect and respond to this kind of important moral good in their environments.

De Waal (2004) suggests a valuable relationship hypothesis to account for reconciliation after aggressive encounters between group members in ongoing relationships that it would be costly to dissolve. How much of a proximate cause recognizing the value of one's ongoing relationships proves to be will again depend on the cognitive sophistication of the species involved, but in humans conscious recognition of the value of our mutually interdependent relationships is certainly part of reconciliation and forgiveness. This point raises the philosophical spectre of partiality as the "real" explanation of this kind of (apparently) moral behaviour, either because reconciliation is selfishly good for the individuals immediately involved or for closely related or reciprocating organisms. To more fully respond to this sort of concern, we will need to discuss, in later parts of the book, attachment and bonding, as well as the relationship between impartiality and moral forms of partiality. Partiality towards those closer to one can be a source of unfair forms of bias, but it can also be a morally good thing, within certain boundaries. We will discuss these boundaries in Chapter 7.

In one highly developed primate, humans, reconciliation is tied to forgiveness and to other important moral goods. Govier (1998, chapters 9 and 10, 2002; chapter 8, 2006; chapters 1, 4, and 5) offers a careful and well-developed analysis of the connections between reconciliation and forgiveness at the individual and group levels, pointing out that while either can occur without the other, cooperative relationships cannot be reliably sustained without both of them. The need for reconciliation and forgiveness arises when one party, the perpetrator, does not take seriously harms that that perpetrator has inflicted upon a second party, the victim. The perpetrator has crossed a boundary relative to the interests of another that he, she, or it should not have. Undoing this boundary crossing in a way that results in sustainable and peaceful cooperation requires a number of things important to our discussion here.

One is acknowledgement that a wrong was done, and the second is an apology for having done this wrong. Both of these acts recognize the moral standing of the other party, something the original wrongful act threatened to obliterate. Also important is genuine remorse and sorrow, emotions that indicate empathy with, and again moral respect with regard to, the former victim. Asking for forgiveness serves to turn the tables, reversing the perpetrator and victim roles by explicitly recognizing that the former victim is now in a position of relative power in terms of rebuilding the relationship through accepting or not accepting the apology that is offered. The apology and its acceptance makes possible the rebuilding of trust and, with trust, sustainable cooperation between the two parties. We will discuss empathy and trust in the next section, two particularly important forms of moral good that start near the beginning of moral trajectories and become more developed as the species on the trajectories become more cognitively and emotionally complex.

More complex moral behaviours are developmentally related to their simpler versions in ways that are mutually supportive. So it is not necessarily the case that moral attitudes or behaviours across different species of social and intelligent animals are the same (moral) kinds of attitudes and behaviours. Rather, these attitudes and the behaviours respond to the same kinds of morally good and bad things in different kinds of cooperative environments. The attitudes and behaviours may be built and operate in different ways, depending on the physiological developmental pathways open to the species in question, and so to that extent they may be different kinds of attitudes and behaviours. But insofar as they respond to moral values, good or bad, they are moral attitudes and behaviours. The behavioural mechanisms involved, however they are constructed, are looking for the same kinds of good or bad things in the organism's environment for the purpose of pursuing or avoiding these things.

Another possible moral good closely connected to the moral goods we have been discussing in this section is altruistic punishment, where a cheater may be punished, for example, by another individual who does not directly gain from this otherwise costly action. When an organism doesn't behave within the moral boundaries it ought to, altruistic punishment provides an important kind of normative push to do so. Again, there is a large and growing literature on altruistic punishment, starting from very basic forms of such punishment and extending to more sophisticated forms of human altruistic punishment, such as the form of altruistic punishment that emerges in situations like Ultimatum games (Fehr and Gachter 2002; Fehr, Fischbacher, and Gachter 2002; Fehr and Fischbacher 2004; Gintis et al. 2005; Gintis 2008; Gospic et al. 2011; Raihani and McAuliffe 2012). Altruistic punishment is connected to fairness and cheating, and its forms grow in complexity in accord with the developing behavioural and psychological sophistication of the species it occurs in. As with other basic moral values, increasing forms of complexity bring with them increasing levels of moral force.

Altruistic punishment is itself a highly complex phenomenon, the moral importance of which the empirical literature is just beginning to explore (see, for example, de Waal (1996, 129–132) for a particularly interesting discussion of this phenomenon among chimpanzees). We conclude this section with a case of punishment gone wrong from de Waal (1996, 91–92). In chimpanzee groups, the alpha male is responsible for policing transgressive behaviour between group members, and in this role he may mete out escalating forms of punishment against transgressors. In mediating transgressive encounters, the alpha male typically favours the underdog in the encounter, even at the expense of his own alliances with other males in the group. But in enforcing group rules, an alpha male may also encounter transgressors who are in the process of transgressing against him. In a case observed by de Waal, an adolescent male chimp was mating with a female the alpha male had previously tried to mate with and been rebuffed by. The subsequent mating was a violation of the group's rules and a punishable offense; and indeed, the female and the younger male had tried to hide their activities from the alpha male, albeit unsuccessfully. The alpha male's pursuit of the younger male was particularly ferocious and intense, with the younger male screaming and defecating in fear as the two chimpanzees raced around the group's enclosure. The alpha male's attacks were challenged and halted by the concerted warning barks of the older females in the group, acting together against the alpha male. So although punishment might be a fair response to something that threatens to upset group relationships and rules, it can also violate its own fair boundaries and become an unfair event in and of itself, that is, a morally bad thing that requires a further moral response of its own.

In all of these cases, what makes mechanisms that detect moral values moral mechanisms is the fact that they detect moral values. Different organisms will develop different mechanisms for doing this, varying from the very simple to the highly complex. What makes them the same kind of mechanism isn't something directly about them but about what they are focused on in the environment with what behavioural result. Mechanisms that take moral values as an input and produce moral consequences as an output are moral mechanisms, however short the circuit between the input and the output might be.

Finally, we note that what is morally right cannot always be directly tied to some particular form of the moral good. Positive moral values may come into conflict, and the more complex they become, the more likely potential conflicts between or among them become. This is especially true for humans, where highly nuanced moral norms are socially articulated and implemented. We will pursue the human case further, and the additional sorts of normativity it involves, in later chapters.

Empathy and trust

In our discussion so far, we have said nothing directly about empathetic caring and trust, two central moral values closely connected to one another and to the other moral values we have been considering. Drawing on the kind of empirical work on empathy represented by Hoffman (2000), de Waal (2009), de Waal and Ferrari (2012), Keysers (2011), and Churchland (2011), we can begin to trace the evolutionary outlines of empathetic caring as it occurs in human emotional and cognitive development, in related animal behaviours, and in the related neurophysiological causes of what appear to be otherwise related behaviours in humans and other animals. This work currently takes us from the involuntary effects of mirror neurons and brain chemicals like oxytocin (Keysers and Churchland) to complex cognitive and motivational capacities like being able to take and act upon the perspectives of other subjects of conscious experience like ourselves (Hoffman).

Like others working in this area, we take these capacities to be moral capacities, with three important differences. Unlike others, we do not take these moral capacities to be the basis of morality. Instead, we think that moral capacities like empathetic caring enable us to recognize and respond to simpler moral values, like helping others who need help. This itself is the second important difference: what makes capacities moral capacities is what they are aimed at. So we are less interested than de Waal and Ferrari, for example, in distinguishing between actual (human) moral capacities and the primate "building block" or "proto" moral capacities that may lead to the more fully developed (human) moral capacities. There is an empirical difference, and an important one; but this difference is not what distinguishes moral from non-moral capacities. Third, because of the ways in which all the capacities in question enable us to become increasingly aware of and moved by shared sets of interests, we think moral capacities arise as moral goods in and of themselves, values to be pursued and encouraged because of the kind of goods that they are. Later moral goods develop from earlier and simpler moral goods by reinforcing and complementing them, as we began to see in the preceding section, and this is what has also happened with regard to empathetic caring and trust.

In terms of their own internal structure, capacities like empathetic caring and trust are normative, insofar as they guide behaviour, and morally normative, insofar as they guide behaviour in moral directions. But what makes these directions moral directions are the moral values that the capacities are aimed at. This matters for the justification of morality, because it is the underlying values, on our view, that ground moral justification, not the normative capacities that have evolved as part of the evolution of moral values. The capacities enable us to think that certain things are morally good, but this does not make them so. What makes moral goods moral goods is that they are naturally arising goods of a particular kind.

On our view, these moral goods are where moral justifications ultimately come to an end. For humans, at our point on our moral trajectory, this is also where moral normativity comes to an end: with moral justifications, based ultimately on natural moral values. We are not assuming that the justificatory pathways from particular beliefs to natural moral goods will be short ones. In our chapter on wide reflective equilibrium, we will argue that moral beliefs are not measured against moral reality one belief at a time, even for simple moral beliefs like "trust is a good thing." Trust among thieves, for example, is morally problematic, and so the claim that "trust is a moral good," as we are stating it here, needs to be taken as more holistically tracking moral truth rather than as stating a particular moral truth. While we think that morally simple beliefs like this can be important in the context of moral arguments, we would also argue that their use is limited in adjudicating among the more complex and conflicting sorts of moral claims that arise for creatures like us, who are able to articulate increasingly complicated moral norms embedded in increasingly complicated social arrangements.

Even so, the crystalizing role that simple claims like "trust is good" can play in moral argument and social change is not insignificant, as we will endeavour to show in our chapter on the abolition of slavery. We will only be in a position to fully address these points about the justification of our moral beliefs in our final chapter, where we discuss moving from the "is" world of natural moral values to the "ought" world of moral justification. The philosophical questions with which we began this chapter are at this point in our argument far from answered.

For now, we want to return to a suggestion we made near the beginning of this chapter: taking a positive interest in the interests of others is a fundamental kind of moral good. Empathetic caring is one significant way of taking a positive interest in the interests of others. As such, it is a morally good kind of thing, directly linked to and reinforcing the more fundamental moral good of taking a positive interest in the interests of others, a moral good which can arise earlier on the moral trajectory that empathetic caring is on, so early, perhaps, that talk of interests may itself be premature. Early forms of empathetic caring arise with the mammalian brain, perhaps with mirror neurons, and with brain chemicals like oxytocin and its neural receptors. Empathetic caring allows one organism to register and respond to positive or negative internal states of another organism, and mirror neurons may be one pathway that enables this to happen, in ways that might or might not involve conscious awareness.

In registering and responding to what happens to and for others, for better or worse, mirror neurons may be one early and important brain mechanism for blurring the distinction between self and others. Electrodes enable us to study mirror neurons in monkeys, and fMRI machines, while much less precise at the neuronal level, enable us to observe what seem to be similar mechanisms in human brains. In terms of actions, feelings, and sensations, mirror neurons enable one individual's brain to represent what is happening to another individual using the same mechanisms for processing what would otherwise be happening to it (Keysers 2011). When one monkey watches another grasp a piece of fruit, the neurons that are activated in the brain of the observing monkey are same ones that are activated when it grasps a piece of fruit. As Keysers puts the point,

> The classical divide between self and other and body and mind becomes fuzzy and permeable in this process. The *mind*-function of predicting another's behaviour [e.g., that a piece of fruit will be grasped and eaten] is now based on the neural representation of the observer's own *body* and actions. . . . The other organism is thus represented in parts of the brain of the observer that were thought to be dedicated to dealing with the monkey's self.
>
> [32]

A similar phenomenon seems to exist for sharing emotions and sensations with others. For example, if you watch someone drinking through a straw make a disgusting face, chances are good that you yourself will experience disgust. Again from Keysers,

> Interestingly, activations in the premotor cortex predicted those in the insula more than the other way around, suggesting that our brain first simulates what the other person's face is doing in the premotor cortex, and once you share the facial expression in your premotor cortex, your insula kicks in, making you share the feelings of that [other] person.
>
> [114]

Focusing on another important aspect of the mammalian brain, Churchland (2011, chapter 3) addresses the same question of how caring about the self is connected to caring about others at the level of neuroendocrinology. Her first question is how brains care about anything, such as what happens to the organism itself, for better or worse. Like Keysers, Churchland's main line of argument is that the same mechanisms that enabled organisms to care about themselves evolved to enable organisms to care about others. Doing things that are good for us releases chemicals like oxytocin in the brain, which then stimulates receptors in the brain in ways that upset the brain's basic equilibrium – having received some oxytocin, the brain responds by trying to produce more. Churchland argues that the evolution of maternal attachment and bonding to offspring harnessed these same chemical processes not just to what might be happening to the mother, as an individual organism, but as well to what might be happening to her offspring. Helping her offspring produces oxytocin in the same ways that helping herself does, so just as a mother helps herself, she helps her offspring. Once this mechanism is in place, it is available, in terms of further evolutionary development, for attachments to others, such as mates or other individuals in a group that might benefit from longer term social bonding. One way in which this might happen is through cooperative breeding and alloparenting, as suggested by Hrdy (2009). Neuroendocrinology makes empathetic caring possible, and empathetic caring makes trust possible. Both make possible longer term social relationships.

In humans, empathetic caring is more psychologically complex than in other mammals. Hoffman (2000) begins his developmental study of human empathy with the reactive cries of newborns, the spark of human empathy that enables humans "to involuntarily and forcefully experience another's emotion," tracing out the development of this capacity to later forms such as veridical empathetic distress and to guilt and the moral internalization of the motive to consider the interests of others (3–9). In later stages of moral development, the motive to consider others is experienced as arising intentionally within the agent himself or herself and as compelling and obligatory (9). Intermediate stages of empathetic development include several egocentric forms of empathetic distress, "in which children respond to another's distress as though they themselves were in distress" (6). At later stages of development, a similar problem can occur with empathic over-arousal, in which an agent's empathy for the distress of another is transformed into an intense feeling of personal distress, causing the agent to become so preoccupied with his or her own distress that the agent's attention is deflected from the distress of the other. So, even if empathy blurs the distinction between self and other, it does not obliterate it, at least in humans. It exists in egocentric forms, but not all its forms are egocentric.

Something similar may be true for other non-human primates as well. Clay and de Waal (2013) observed what appeared to be emphatic over-arousal in bonobos that had been orphaned at a young age due to the illegal bushmeat and pet trades in certain areas of Africa. Compared to bonobos that had been reared by their own mothers, the developmentally delayed bonobos showed significantly less spontaneous consolation behaviour after naturally occurring aggressive or otherwise stressful encounters. Bonobo juveniles reared by their own mothers were likely to recover more quickly from their own distress when victimized by such encounters and more likely to console other victims when they were third-party observers of such encounters. Juvenile bonobos that were orphaned were more likely to flee such encounters.

An important question for those working on non-human forms of empathy (see, for example, de Waal and Ferrari (2010) and de Waal and Ferrari (2012)) is what makes different forms of empathetic caring the same kind of moral or at least proto-moral capacity as the human capacity. The quickest answer is that they are aimed at the same kinds of things in the same kind of way. For us, this is precisely what makes them moral capacities: they are aimed at moral goods. Because of how they are intertwined with these goods, they are moral goods themselves. Empathetic caring is itself a good kind of thing, one which arises on evolutionary trajectories that start from simpler good kinds of things. There is no reason to suppose our current human form of empathy is an endpoint on our own trajectory: its presence together with future evolutionary pressures may push our trajectory in more complex moral directions. There may be moral values we are unaware of or aspects of current moral values that we are unaware of. We know more about empathetic caring than bonobos, but other species might develop that could know more about empathetic caring and morality than we do. It may be easy for us to see the limits of bonobo's moral knowledge but much harder for us to see the limits of our own. Species more morally developed than us might not have this same moral problem.

Here we might usefully return to the experiment from Chapter 2 that involved the monkeys who stop pulling the chain that delivers food to them when they notice it also delivers a shock to the monkey in the cage next to them. When it comes to empathy, monkeys are not bonobos, nor are they rats. But as Church (1959) discovered, rats will also stop pushing a bar that delivers food to them if they notice that pushing the bar also shocks a rat in the cage next to them. Perhaps this is an empathetic response, perhaps it isn't; again, one wants to know what is going on in the rats' brains, as with the monkeys'. Maybe all that is going on is emotional contagion, and the rats can stop their own distress by refraining from pushing the bar.

As we have already noted, juvenile rats engage in fair rough-and-tumble play, and as well-socialized adults competing for rank and resources they do also, in appropriate circumstances, aggressively harm one another. Apparently, they can distinguish their distress from the distress of others, and when the distress of others can be tolerated and when it cannot. So perhaps in the bar-pushing experiment, successfully socialized rats simply notice that they are harming another rat when they should not be. So maybe empathy is not what explains the behaviour but instead just knowing the boundaries of when to harm and when not to harm another rat.

On the other hand, rats will help a distressed fellow rat rather than simply open a door and eat the chocolate behind it. Bartal, Decety, and Mason (2011) put free rats into a cage with a restraining container the door of which could be opened with practice. Once they had learned how, rats reliably freed restrained rats, and when given a choice between freeing a restrained rat and opening another container that contained food, they opened both doors and often shared the food. So perhaps when they are delivering a shock, what they are noticing is a rat that somehow needs help.

Our point here is not to argue for or against any of these empirical hypotheses but to point out that the environment of rats is full of moral goods, whether their brains and minds are evolved enough, yet, to notice and respond to any or all of these goods in the right sorts of ways. Even if rats were unable to stop pushing the bar when another rat needed help or was being shocked, it would be a morally good thing if they were. Maybe a species on a moral trajectory is able to notice this, maybe not. It would be a morally good thing if they could.

Having moved from rats, monkeys, and apes to humans, we end this section with two animal studies that link the boundaries of fair play with emotional contagion and potentially with empathy and trust. Ross, Menzier, and Zimmerman (2008) studied rough-and-tumble play in orangutans and discovered significant levels of rapid facial mimicry between playmates. In humans, rapid facial mimicry seems to be tied to empathy, mediated, as we saw in Keysers' discussion, by mirror neurons. In rough-and-tumble play, being able to read the emotions of the other may be part of trusting that the playful encounter is just that. Mancini, Ferrari, and Palagi (2013) found rapid facial mimicry in Gelada monkeys, also in the context of rough-and-tumble play. Just as empathy and trust emerge developmentally in humans, they also seem to emerge phylogenetically as species become more socially intelligent. Such species are, we think, on moral trajectories that make the emergence of more complex forms of moral goodness possible. Morality lies along such trajectories rather than simply at one particular point on one particular trajectory. Humans are not the sole measure of morality.

Impartiality

The biological importance of empathy and trust might appear to raise a significant objection to any attempt to base ethics on evolution. Steven Mithen (2012) raises the objection evocatively in his review of E.O. Wilson's (2012) *The Social Conquest of Earth*, a chapter of which is entitled "Tribalism Is a Fundamental Human Trait." At its simplest, the objection is that any morality that evolved to favour in-group individuals over out-group individuals is morally bankrupt at its core.

The general worry is that any biologically based morality that can be traced back to such things as attachment and bonding is going to be fundamentally partial at its core rather than impartial. Although we think that some moral values are, importantly, partial, we also think that morality is more fundamentally impartial than partial.

In his book *The Expanding Circle*, Peter Singer (2011, chapter 4) solves the potential problem of moral partiality by an appeal to human reason. If I want my interests to count as worthy of the respect of others, then to be consistent, I must also allow the interests of others to count as being worthy of my own respect. That certain interests are mine or my group's is not a morally relevant feature of those interests, according to Singer's argument. For us, however, reason is too late on the scene to create morality.

There is also the deeper question of how my own interests become the sort of thing worthy of the respect of others. At its deepest level, is morality just a matter of my own wants? Satisfied interests are generally good things for the individuals who possess them, but what makes them morally good, in and of themselves? How does morality enter into this argument at the ground level, in such a way that reason can take over and give moral worth to the interests of an ever-expanding set of creatures whose interests are worthy of our human respect?

If we are utilitarians, like Singer, we could say that happiness is good thing, and so the more of it the better. But for us, happiness per se is not a moral good, even if being happy is certainly better, as a general rule, than being unhappy. One might say the same thing, of course, about being rich or poor: if you've experienced both, being rich is clearly better. On our view, happiness and satisfied interests, all by themselves, are non-moral goods.

We think that from a biological point of view, the most fundamental moral good is taking a positive interest in the interests (or good) of another. This means that at its deepest level, morality is fundamentally impartial and impartiality is fundamentally impersonal. The form of impartiality at the core of morality focuses on interests, or things like interests, but certainly not persons. Persons are not around at the beginning of morality, and the distinction between self and other is blurry at best. The question of how best to understand human impartiality takes us to a deep divide between Kantians, like John Rawls, and utilitarians. A central claim made by Rawls (1971, 27) is that utilitarians, in maximizing the total number of interests satisfied regardless of how those interests are distributed across individual lives, fail to take seriously the separateness of persons. The impartiality that is built into Rawls' Original Position is meant to provide a competing conception of impartiality that is not impersonal in this sort of way. The Original Position treats all persons as moral equals, in a way that utilitarianism does not.

So despite our comments on Singer, we are suggesting that a utilitarian form of impartiality is at the core of morality. On the other hand, by tracing morality back to the earliest cases where the interests of another organism count in a positive way for an organism that is at the beginning of a moral trajectory, we do need to acknowledge that not all the interests of all others are initially important to the individual organism in question. This means that from the outset of morality, partial moral concerns exist in tension with impartial concerns. Certain interests of specific others matter to the individual organism, not all the interests of all others.

Still, when the individual organism is moved by the interests of individuals it is related or at least in close proximity to, it is moved by those interests as interests that matter to it. This is the beginning of the impersonal form of impartiality. We think that this impartiality is a basic form of moral goodness.

For organisms that can think, talk, and reason, negotiating the tension between partial and impartial moral concerns will be a source of ongoing argument and adjustment. We will need to explore some of the dimensions of this kind of argument and adjustment in our later chapters on the abolition of slavery and on wide reflective equilibrium. In defence of Rawls' Kantian conception of impartiality, we should also remember that when autonomy appears on a moral trajectory, respect for the autonomous choices of others becomes possible as a more complex form of moral value. On an evolutionary approach to ethics, this poses an interesting problem, because utilitarian and Kantian forms of impartiality are not reducible to one another.

For organisms that cannot think, talk, and reason, the tensions between self and different sorts of others will exist as pressures within their social environments that will need to be carefully negotiated, depending on the level of behavioural flexibility of each particular species that is on a moral trajectory. For species with less flexibility, more direct evolutionary pressures will sort out the environmental tensions as well as they can be sorted out for the species in question. For organisms that are not human, partial moral concerns will more often than not trump impartial concerns. But not always: some apes will drown in moats as they try to save human toddlers not carefully attended to by their parents.

Bibliography

Anderson, James R., Hika Kuroshima, Ayaka Takimoto, and Kazuo Fujita. 2013. "Third-Party Social Evaluation of Humans by Monkeys." *Nature Communications* 4. doi: 10.1038ncomms2495.

Anderson, James R., Ayaka Takimoto, Hika Kuroshima, and Kazuo Fujita. 2013. "Capuchin Monkeys Judge Third-Party Reciprocity." *Cognition* 127:140–146.

Aureli, Filippo, and Frans B.M. de Waal. 2000. *Natural Conflict Resolution*. Berkeley: University of California Press.

Bartal, Inbal Ben-Ami, Jean Decety, and Peggy Mason. 2011. "Empathy and Pro-Social Behaviour in Rats." *Science* 334:1427–1430.

Church, R.M. 1959. "Emotional Reactions of Rats to the Pain of Others." *Journal of Comparative and Physiological Psychology* 52 (2):132–134.

Churchland, Patricia. 2011. *Braintrust: What Neuroscience Tells Us about Morality*. Princeton: Princeton University Press.

Clay, Zanna, and Frans B.M. de Waal. 2013. "Development of Socio-Emotional Competence in Bonobos." *PNAS* 110 (45): 18121–18126.

Cools, Annemieke K.A., Alain J.-M. Van Hout, and Mark H.J. Nelissen. 2008. "Canine Reconciliation and Third-Party Initiated Postconflict Affiliation: Do Peacemaking Social Mechanisms in Dogs Rival Those of Higher Primates?" *Ethology* 114 (1):53–63.

de Waal, Frans B.M. 1996. *Good Natured: The Origins of Right and Wrong in Humans and Other Animals*. Cambridge, MA: Harvard University Press.

de Waal, Frans B.M. 2000. "Primates: A Natural Heritage of Conflict Resolution." *Science* 289 (5479):586–590.
de Waal, Frans B.M. 2004. "Evolutionary Ethics, Aggression, and Violence: Lessons from Primate Research." *Journal of Law, Medicine & Ethics* 32:8–23.
de Waal, Frans B.M. 2009. *The Age of Empathy: Nature's Lessons for a Kinder Society*. New York: Random House.
de Waal, Frans B.M., and Pier Francesco Ferrari. 2010. "Toward a Bottom-Up Perspective on Animal and Human Cognition." *Trends in Cognitive Sciences* 14:201–207.
de Waal, Frans B.M., and Pier Francesco Ferrari, eds. 2012. *The Primate Mind: Built to Connect with Other Minds*. Cambridge, MA: Harvard University Press.
Dewey, John, and James H. Tufts. 1908. *Ethics*. New York: Henry Holt & Co.
Fehr, Ernest, and Urs Fischbacher. 2004. "Third-Party Punishment and Social Norms." *Evolution and Human Behaviour* 25 (2):63–87.
Fehr, Ernest, Urs Fischbacher, and Simon Gachter. 2002. "Strong Reciprocity, Human Cooperation, and the Enforcement of Social Norms." *Human Nature* 13 (1):1–25.
Fehr, Ernest, and Simon Gachter. 2002. "Altruistic Punishment in Humans." *Nature* 415: 137–140.
Gintis, Herbert. 2008. "Punishment and Cooperation." *Science* 319:1345–1346.
Gintis, Herbert, Samuel Bowles, Robert Boyd, and Ernest Fehr, eds. 2005. *Moral Sentiments and Material Interests: The Foundations of Cooperation in Economic Life*. Edited by Ken Binmore, *Economic Learning and Social Evolution*. Cambridge, MA: MIT Press.
Gospic, Katarina, Erik Mohlin, Peter Fransson, Predrag Petrovic, Magnus Johannesson, and Martin Ingvar. 2011. "Limbic Justice: Amygdala Involvement in Immediate Rejection in the Ultimatum Game." *PLoS Biology* 9 (5):e1001054. doi: 10.1371/journal.pbio.1001054.
Govier, Trudy. 1998. *Dilemmas of Trust*. Montreal and Kingston: McGill-Queen's University Press.
Govier, Trudy. 2002. *Forgiveness and Revenge*. New York: Routledge.
Govier, Trudy. 2006. *Taking Wrongs Seriously: Acknowledgement, Reconciliation, and the Politics of Sustainable Peace*. Amherst and New York: Humanity Books.
Gueguen, Nicolas. 2012. "The Sweet Smell of . . . Implicit Helping: Effects of Pleasant Ambient Fragrance on Spontaneous Help in Shopping Malls." *The Journal of Social Psychology* 152 (4):397–400.
Hamlin, J. Kiley, and Karen Wynn. 2011. "Young Infants Prefer Prosocial to Antisocial Others." *Cognitive Development* 26 (1):30–39.
Hamlin, J. Kiley, Karen Wynn, and Paul Bloom. 2007. "Social Evaluation by Preverbal Infants." *Nature* 450:557–559.
Hoffman, Martin L. 2000. *Empathy and Moral Development: Implications for Caring and Justice*. Cambridge: Cambridge University Press.
Hrdy, Sarah Blaffer. 2009. *Mothers and Others: The Evolutionary Origins of Mutual Understanding*. Cambridge, MA: Harvard University Press.
Keysers, Christian. 2011. *The Empathetic Brain: How the Discovery of Mirror Neurons Changes Our Understanding of Human Nature*. Amazon: Kindle E-Book.
Mackie, John Leslie. 1977. *Ethics: Inventing Right and Wrong*. Harmondsworth: Penguin.
Mancini, Giada, Pier Francesco Ferrari, and Elisabetta Palagi. 2013. "Rapid Facial Mimicry in Geladas." *Scientific Reports* 3:1–5. doi: 10.1038/srep01527.
Mithen, Steven. 2012. "How Fit Is E.O. Wilson's Evolution?" *The New York Review of Books* 59 (11):26–28.

Moore, G.E. 1903. *Principia Ethica*. Cambridge: Cambridge University Press.
Pellis, Sergio M., and Vivien C. Pellis. 1996. "On Knowing It's Only Play: The Role of Play Signals in Play Fighting." *Aggression and Violent Behavior* 1 (3):249–268.
Pellis, Sergio M., and Vivien C. Pellis. 2009. *The Playful Brain: Venturing to the Limits of Neuroscience*. London: Oneworld.
Pellis, Sergio M., Vivien C. Pellis, and C.J. Reinhart. 2010. "The Evolution of Social Play." In *Formative Experiences: The Interaction of Caregiving, Culture and Developmental Psychobiology*, edited by C. Worthman, P. Plotsky, Schechter D. and C. Cummings. Cambridge: Cambridge University Press.
Raihani, N.J., and Katherine McAuliffe. 2012. "Human Punishment Is Motivated by Inequity Aversion, Not a Desire for Reciprocity." *Biology Letters*. doi: 10.1098/rsbl.2012.0470.
Rawls, John. 1971. *A Theory of Justice*. Cambridge, MA: Harvard University Press.
Ross, Marina Davila, Susanne Menzier, and Elke Zimmerman. 2008. "Rapid Facial Mimicry in Orangutan Play." *Biology Letters* 4 (1):27–30.
Seed, Amanda M., Nicola S. Clayton, and Nathan J. Emory. 2007. "Postconflict Third-Party Affiliation in Rooks, Corvus Frugilegus." *Current Biology* 17 (2):152–158.
Singer, Peter. 2011. *The Expanding Circle: Ethics, Evolution, and Moral Progress*. 2nd ed. Princeton: Princeton University Press.
Stingl, Michael. 1997. "Ethics I (1900–45)." In *Philosophy of Meaning, Knowledge and Value in the 20th Century*, edited by John V. Canfield, 134–162. London: Routledge.
Warneken, Felix, Brian Hare, Alida P. Melis, Daniel Hanus, and Michael Tomasello. 2007. "Spontaneous Altruism by Chimpanzees and Young Children." *PLoS Biology* 5 (7):e184. https://doi.org/10.1371/journal.pbio.0050184.
Warneken, Felix, and Michael Tomasello. 2006. "Altruistic Helping in Human Infants and Young Chimpanzees." *Science* 311:1301–1303.
Warneken, Felix, and Michael Tomasello. 2007. "Helping and Cooperation at 14 Months of Age." *Infancy* 11 (3):271–294.
Wilson, Edward O. 2012. *The Social Conquest of Earth*. New York: Liveright.

4 Moral sense theories

Moral sense theories and moral truth

Since we began thinking about EMR in the early 1990s, there has been an explosion of empirical work in the developmental and comparative psychology of moral cognition. Moral sense theories are especially important if we suppose that morality itself may have an empirical basis in organisms' environments. Such theories are thus central to the project of this book. The fact that the human moral sense has a biological structure suggests that whatever morality turns out to be, it cannot be simply reduced to social contracts based on rational self-interest or rational self-interest coupled with equal respect for others – two of the main naturalistic theories of moral truth we will examine in our next chapter. EMR includes the idea that the human moral sense and contractualist thinking are both important components of morality as it has come to exist in humans.

EMR claims, additionally, that moral values themselves are important parts of the world as it exists in itself, apart from human thought, perceptions, and emotions. The moral (or proto-moral) senses of different species, including and perhaps limited to species of the *Homo* lineage, are certainly real features of the natural biological world. By themselves they do not, however, give us a biological version of moral realism or a firm foundation for a biologically based theory of moral truth. If our human moral sense does nothing more than create particularly human moral values rather than also enabling us to discover them, a version of evolutionary moral naturalism is true but not a version of EMR. Moral value would ultimately depend on the perceivers of such value, not the other way around.

The central empirical claim of EMR is that the human moral sense and the related pro-social capacities of other animals are selected for by generally occurring structural features of certain kinds of environments, features that can be identified and studied as the distinct biological kinds of things that they are. The second half of the book argues that these kinds are moral kinds, and thus all responses to them of the appropriate kind are moral responses. This chapter continues our argument that supposing such kinds to exist has explanatory value in the study of evolutionary biology.

Moral instincts

From the point of view of moral sense theories, having a moral sense may be best understood as a property of members of the *Homo* genus or perhaps particular branches of this genus insofar as morality requires language, thought, and the sort of well-developed self-determination that underwrites moral responsibility. The human moral sense might be related to earlier and less well-developed cognitive and prosocial emotional capacities of similar kinds but not of the same kind. Because of their developmental link to human moral capacities, we might think of these earlier capacities as proto-moral capacities, whether they were the capacities of our own ancestors or capacities of species at similar levels of evolutionary development. This is essentially the view of Frans de Waal.

A main focus of de Waal's influential research program in evolutionary psychology over the past several decades has been moral instincts. Beginning with de Waal (1996), important publications regarding moral instincts and morality include Flack and de Waal (2000), de Waal (2006, 2009, 2016), and de Waal and Ferrari (2012). De Waal does not suggest that we can reduce morality itself to moral instincts; more precisely, he thinks that moral instincts are necessary but not sufficient for morality. Morality requires thought, argument, and conscious intentions based on thinking and arguing. But according to de Waal, there would be nothing "moral" to think and argue about if we humans did not share with other primates and mammals a basic set of prosocial emotions involving such things as empathy, fairness, consolation, and retribution. These emotions lead us to care about the interests of others and about the common good of the groups we are part of as social and intelligent primates. In this way, morality is, for de Waal, anchored in the real world of biology, but the real aspects of the biological world that morality is based on are all at the instinctual or psychological level of biological explanation.

Moral instincts are an important part of EMR as we are developing it in this book. For us, de Waal's work provides an emerging and detailed account of what moral instincts look like within and across a wide variety of species. As its own independent approach to the biological foundations of morality, this work anchors human moral discourse in prosocial emotions and the patterns of thought and argument that these emotions make possible in the human species. But where de Waal's approach sees the origins of morality in the emotional and cognitive psychological capacities of the species he studies, EMR sees the origins of morality in the structural similarities of the social environments in which these capacities develop. For EMR, moral values begin in common features of social environments that are not specific to the instincts of any particular species. EMR looks not just to instincts but to features of a species' environment that were part of the selection process that produced those instincts.

In one of de Waal's most famous examples, grapes are given to one monkey where other monkeys have been satisfied getting cucumbers in return for performing the same task. The cucumbers are no longer good enough: they are refused, thrown on the ground, or thrown back at the experimenter. The effect is striking, and our attention is easily focused on the monkeys and on their immediate psychological responses to getting the lesser reward while another monkey is receiving a greater reward for doing the same thing they are. We relate to their anger and frustration, but we also see what it is about their situation that is triggering this anger and frustration.

What EMR specifically draws our attention to are the details of the situation that triggers this kind of instinctual response and to the underlying similarities of these details to those of similar situations that arise in the environments of other species. EMR identifies the regularly recurring structural elements of such situations as a natural moral kind and argues that these sorts of natural moral kinds help to explain the sorts of psychological traits observed by de Waal. While de Waal and other comparative psychologists have understandably been focused on the interesting cognitive traits that seem to emerge in the experimental situations under study, EMR shifts our theoretical perspective from these traits to the common and regularly recurring features of the environments these traits arise in response to.

The question is, why take this further theoretical step and suppose that there are natural moral values that moral (or proto-moral) capacities evolve in response to? In this chapter, we are still mainly considering the first half of this question, that pertaining to the ontological existence of the kinds in question. The second half of the question, left for the second half of this book, is why we should think that these biological kinds are moral kinds. So here we are asking, why should we suppose that there are general environmental kinds that explain the evolutionary development of a wide variety of moral or proto-moral instincts? First, the regularly recurring structural features of the environments in question seem to be there, in and across the environments in which the traits develop. Second, their presence in these environments seems to offer avenues of explanation for where and how the traits develop. Third, these avenues of explanation suggest a foundation for a promising new research program.

This third reason is worth pausing to consider. De Waal's more behaviourist critics raise a parallel question for his own basing of morality on psychological capacities like an aversion to unfairness. Why suppose complicated cognitive capacities rather than just certain behavioural patterns triggered in response to certain features of the species' environment? In the last chapter, we gave a general response to this line of thought, but here we reconsider the same problem with regard to the sorts of specific moral instincts hypothesized by de Waal's research program.

A ready target for such criticism comes from an example in de Waal (2016, 53–54) involving chimpanzees and hidden grapefruit. Within the chimps' sight, grapefruit are carried past their nighttime enclosure and onto the island they occupy during the day. The grapefruit are buried on the path onto the island with only bits of skin showing. A more junior male chimp appears to pause at this point on the path as the group of chimps run onto the island. While the other chimps are napping later in the day, this particular chimp returns to the buried grapefruit, unearthing and eating them. Had he alerted others to their whereabouts, he might not have gotten any. De Waal takes this to be an instance of taking into account the perspective of others, an important aspect of chimp cognition and a capacity important for the development of morality.

Against this explanation of the chimps' behaviour, a more behaviourist comparative psychologist might suppose more simply that the chimp noticed something as he was running along the path and then went back later to explore the ground more carefully. Or even more simply, the chimp wasn't particularly tired that day and wandered back to the pathway that the group had traversed earlier. How do we rule out such alternative explanations? Stingl (2016) suggests that the answer to this question will not likely hinge on simply doing the right sort of experiment. Single experiments will always face alternative explanations. Series of experiments can make some of these explanations more or less likely, but more interesting still is the research program that such experiments are embedded in: does the research program in question lead us to interesting experimental results that we otherwise probably would not have gotten? De Waal's approach to evolutionary cognition led him to the grapefruit experiment and to the result produced by this experiment. Would a more behaviourist approach have led to this same experiment and its result? If a research program is significantly productive in this regard, as de Waal's would appear to be, this suggests that the program is on the track of something real.

An important theme of de Waal (2016) is that it is challenging to think up experiments that will test for the particular cognitive capacities of a species because we do not share its species-specific Umwelt: its perspective on what is interesting and important in the world around it. One way to respond to such challenges is to hypothesize instincts that are continuous with our own and then set up experimental situations that predict the sort of behaviour that might be supposed to be produced by the instinct in question. So if we think a species of primate is capable of taking the perspective of another individual and sympathizing with its plight, we might set up an experiment where one individual needs help that another individual is in a position to provide. But here we will need to know what counts as a plight for the species in question and what counts as help. For dolphins, one important kind of plight is struggling near the surface of water for air, and a helpful gesture is to use one's body to help buoy up the struggling individual. For chimps, being too small to swing between two branches is an important kind of plight, and mother chimps will bridge such gaps with their own bodies to allow their children to get across them. What chimps notice, and their capacity to respond to what they notice, is not what dolphins are likely to notice or be able to respond to, and vice versa.

What EMR adds to this research program is the idea that to construct the right sorts of experimental situations, what de Waal and his colleagues are in fact doing is thinking about what it is to help one another in the environments in question. How does this kind of thing, to help another when he or she is struggling, manifest itself in the environment of this particular social and intelligent species? In EMR's terms, how does the moral value of helping one another manifest itself in this species' biological environment? EMR starts the biological explanation of morality with moral values in the environment, not with the instincts that have developed in response to these values. So while de Waal anchors morality to the biological reality of moral instincts, EMR anchors morality to the structurally recurring features of the environment of social and intelligent species that these instincts evolve in response to. Moral values, as natural moral kinds, are a part of explaining why a species' Umwelt has the particular moral structure that it does and also why different species have similar but different Umwelts of the same general kind.

What constellation of features in a species' environment, structurally, could count as helping another? Here EMR suggests that one look past species with potentially well-developed proto-moral capacities to species that are in the right sorts of environments for moral values to appear and to make a difference. Why look for these sorts of structural elements in a species' environment, along with their possible developmental consequences, unless one supposes, along with EMR, that they exist, and that they exist as precisely the kinds of thing that they are? Likewise, in terms of instinctually better-developed species, EMR tells us to look more closely at all the kinds of moral values that might be available in the environment of these species, to see exactly how these kinds of values might or might not be linked to the instinctual capacities of the species in question. The leading question is always how the structures of instinctual responses might be linked to the structural features of the environment that trigger them, or could trigger them, were the species to develop in this direction.

If we then ask of EMR, why count the underlying structural elements of the environment as moral values, the answer to the empirical half of this question is that by exploring moral values as the natural kind of thing that they appear to be, we can generate a more robust research program for testing alternative theories about moral instincts. From the point of view of evolutionary biology, environmentally based natural moral kinds make explanatory sense. From the point of view of moral philosophy, they would also provide a foundation for moral truth, once a species develops the capacity to talk and to argue about them. These arguments would then fundamentally not be about the content of species-specific instincts but about the very real moral values in the specific and more general biological environments that such instincts evolve in response to. This difference matters to the question of moral truth, the topic of our final chapter. Moral values are part of the human biological environment and getting them wrong, or at least wrong enough, can be disastrous for us, in the same general kind of way that getting germs wrong can, as a matter of fact, be disastrous for us.

We return to the question of moral truth and its importance for biological creatures like us in our final chapter. Our point here is that in addition to the reality of moral values as natural moral kinds, EMR is also committed to the reality of moral instincts. To give more empirical content to the idea of moral instincts, we examine several of de Waal's examples.

For de Waal, the capacities for empathy and fairness are the two pillars of morality. De Waal's work includes many examples of empathy but a particularly interesting one involves the Russian primatologist Nadia Ladygina-Kohts who raised a young chimp, Yoni, along with her own young son. We again stress that this is one example among many, and that it is not experimentally produced. De Waal (2009, 86) quotes Ladygina-Kohts at length:

> If I pretend to be crying, close my eyes and weep, Yoni immediately stops his play or any other activities and quickly runs over to me, all excited and shagged, from the most remote places in the house, such as the roof or the ceiling of his cage, from where I could not drive him despite my persistent calls and entreaties. He hastily runs around me, as if looking for the offender; looking at my face, he tenderly takes my chin in his palm, lightly touches my face with his fingers, as though trying to understand what is happening, and turns around, clenching his toes into fists.

From her continued description of this incident, de Waal says that "when she slapped her hands over her eyes, he tried to pull them away, extending his lips toward her face, looking attentively, slightly groaning and whimpering" (86). As de Waal comments, if Yoni was concerned only at his own discomfort, he could have moved himself further away from the crying or left Ladygina-Kohts' hands where they were. "Clearly," says de Waal, "Yoni wasn't just focusing on his own situation: he felt an urge to understand what was the matter with Kohts" (88). As de Waal further notes, if we were talking about Ladygina-Kohts' young son, we would not hesitate to call this sort of response sympathetic, in the sense that sympathy is proactive where empathy more narrowly construed is cognitive. Sympathy, or empathy more broadly construed, involves concern with the other's situation and a desire to do something to make it better. As with helping another who needs it, the instinctual capacity involved here is aimed at a certain kind of environmental feature that can vary with the environments of the species in question. While de Waal and other comparative psychologists are focused on the capacities in question, EMR is focused more broadly on the underlying environmental regularities that these capacities evolve in response to.

Regarding the capacity for fairness, de Waal's nicest example is the one we have mentioned several times, the capuchin monkeys receiving either grapes or cucumbers. Flack and de Waal (2000) add a number of other behaviours and related capacities to the two-pillar approach to morality of empathy and fairness, including reconciliation, consolation, conflict intervention and mediation, community concern, a sense of social regularity and related social expectations, social commitment, social rules and moralistic aggression when these rules are violated, and finally the ability to take into account and to balance the interests of others against one's own interests. De Waal's overall approach to moral instincts thus involves much more than empathy and fairness. We will not give examples here of all these behaviours and capacities, but we do want to mention two recent and interesting studies of a kind that de Waal's work might lead us to expect.

Regarding what we might label as a concern for the common good, or at least the good of the community, Langergraber et al. (2017) provide the example of male chimps regularly patrolling the boundaries of their territory. These patrols often encounter rival patrols creating situations that can lead to aggression and injury or death. Some of the chimps who participate in the patrols have relatives in the group to protect but others do not; however, chimps do better in terms of mating opportunities the larger their groups and defeating rival patrols can lead to larger group size. Bigger patrol groups seem to be a common good that chimps are able to recognize, even though they might benefit from patrolling behaviour without themselves participating.

Regarding social expectations and sharing at a cost to oneself, Schmelz et al. (2017) found that chimps would share food with others at a cost to themselves in circumstances where the other had earlier helped them or had taken a risk that had benefitted them. For example, chimps could choose whether to obtain a larger reward for themselves or smaller rewards for both themselves and another. They could also choose whether to do so, or to let the other chimp choose first. In an experiment where one chimp was trained to defer the choice to the other, the second chimp often picked the joint sharing option.

These are of course only two studies, and there are other possible explanations for the results of both of them. The point remains that the results are surprising and that we would probably not have looked for them outside de Waal's research program. The more such results we are able to generate, the more robust de Waal's research project would appear to be.

In humans, capacities of similar kinds are much more sophisticated, and there is built on top of them language, reasoned argument, and the continuous development of systems of moral rules. There is also the sense that the obligations generated by such rules have a form of validity that goes beyond any emotional commitment we might have to the content of the rules or beyond the mere fact that we have jointly agreed to follow the rules themselves. Stanford (2018) raises this point against de Waal in a well-focused way, and his own argument regarding the external validity of moral reasons and moral obligations provides us with a final point with which to conclude this section.

If we suppose along with de Waal and others that moral emotions are the building blocks of human morality, we face the problem of accounting for the felt externality of our moral obligations. Moral emotions might help to give moral content to some of our social rules, but what gives these rules, as moral rules, their felt obligatory force?

We might of course suppose that this obligatory force is illusory. Maybe we humans, as semi-intelligent primates, are simply prone to projecting our strong emotional responses to objects onto or into the objects themselves, that is, as supposed properties of these objects. Fruit that seems really sweet to us is itself really sweet. But as Stanford notes, the problem with this line of argument is that we do not, on reflection, tend to think that properties like sweetness really are, in and of themselves, actual properties of objects, and we certainly do not do this with one of our strongest emotional responses to some objects, namely the ones that cause us pain. The pain is in us, not in the object.

Joyce (2006, 123–133) offers a more targeted version of an error theory approach, arguing that moral values are special: if we humans did not project moral properties onto objects, we would not think moral rules obligatory. And if we did not think of moral rules as obligatory, such rules would not override the demands of self-interest often enough for us to succeed in socially cooperative groups. But as Stanford responds, if our biological interests hinge on recognizing the force of moral rules, they hinge on other values as well, so why a set of human psychological proclivities that make moral values trump cards over all other values, all the time, or as much of the time as possible? Why not just make the pull of moral values particularly strong, like the pull of pain, which is, again, entirely subjective?

Stanford's own view is that to protect ourselves from cheaters in our groups, either others or ourselves, we needed to feel strongly that the social rules that we were developing applied to everyone in the group equally, ourselves as well as others. What we were strongly inclined to look for were others in our group who shared exactly this same attitude, as strongly as we ourselves did. If this view is right, the felt external validity of moral rules comes at least in part from the human need to associate with others who share our sense of obligation to rules we have together agreed upon. If this is right, Stanford's view provides an empirical link between EMR and the contractualist views we will consider in our next chapter by reinforcing, in the human case, the externality of the biologically based natural moral values hypothesized by EMR.

Unlike other species, humans are reflectively aware of moral values, as well as the conflicts between moral values and other values. As we reflect and argue about such conflicts, the need to associate with others who share our moral attitudes adds force to the moral values that already exist externally to us in our social environment. Social contracts are then supported by the natural moral values at their base, plus the need to associate with others who share our moral attitudes, plus the rational commitment to respect the conclusions of the moral arguments that we enter into with one another. That is, humans are also rational, and rationality itself provides some of the binding power of the rational contracts postulated by the contractualists, contracts which exist externally to each of us who might be rationally supposed to be part of any such contractual arrangement. The externality of moral rules thus has several sources, only one of which Stanford identifies, which is an intermediate one at that.

On Stanford's view, it is not surprising that the social rules we regard as obligatory often involve content centred on such moral values as playing fair or not harming others. Rules aimed at such values can be expected to be especially cohesive in human groups where harming others or giving them less than an expected share is apt to lead to fighting. EMR makes this unsurprising feature of Stanford's approach even less surprising by generalizing the point to apply it beyond hominins to hominoids to any species that is both intelligent and social. The moral content of the likely rules of human moral systems and human morality more broadly is not simply due to a quirk of the human species and its social environment.

To sum up our discussion here, EMR adds a significant element to de Waal's approach to moral instincts and to arguments like Stanford's regarding the external force of moral obligations in humans. If Stanford is right, his approach provides a stepping stone from instinctual responses to the moral values hypothesized to exist by EMR to the social contracts of theorists like Rawls, Scanlon, and Copp discussed in the next chapter. Abiding by social contracts based on a combination of rationality and impartiality appeals as deeply as it does to us because of the deeply felt need to hold ourselves and others equally accountable to the social rules we have decided upon to regulate the groups that we are parts of.

Given our reflectively developed sense of conflict between moral and non-moral values, the pull of natural moral values might not succeed enough of the time to enable humans to succeed in social groups. But the pull may still be there: for humans, like capuchins, the pull of fairness itself may be biologically real, and fairness requires us to treat the same cases in the same way. If the rules apply to me, they apply to you as well, and vice versa. Nor should we forget the force of rational argument and agreement, particularly in a context where we are engaged in reasoning together about what we together ought to do. Treating like cases alike is also a commitment of rationality, as is accepting conclusions that are supported by premises we jointly accept. Our point here is that there are a number of possible sources for the felt external validity of moral obligations, and while EMR provides a beginning point for such additional sources to build upon, it also helps to explain more fully why felt externality may be an indicator of real externality. The externality of moral values does not begin (or necessarily) end with the evolution of the human species and the sort of psychological capacities we ourselves needed to succeed and flourish in social groups as intelligent primates reflectively aware of our own interests. For EMR, moral normativity may be a multilayered phenomenon, depending on where a species may be on its particular moral trajectory.

Maternal care and alloparenting

For EMR, Sarah Blaffer Hrdy (2009) is interesting because of its focus on the evolutionary development of the *Homo* genus, asking how we developed our particularly robust combination of intelligence and prosocial emotions. Unlike other apes, humans are cooperative breeders, where mothers readily share their offspring with alloparents: grandmothers, aunts, and adult males within their immediate social group. No other apes are cooperative breeders, and it is Hrdy's view that our human relationship to morality is grounded both in our higher intelligence as apes and the fact that as a genus and species of apes we are cooperative breeders. For Hrdy, our well-developed capacities for a linked set of moral emotions – sharing, trust, and empathy chief among them – predated language and our capacity to articulate and argue about moral judgments and rules.

Hrdy (2009) addresses two main questions. First, why, among hominids, did the *Homo* genus emerge as cooperative breeders? What was special about our environment? Second, how, in highly intelligent hominins, did the prosocial emotions develop as powerful forces of cooperative breeding, both for alloparents and for the infants and young children of alloparents? What emotional capacities would enable a *Homo* mother to pass her infants to others, what would enable these others to offer extended forms of parental care to these infants, and what about the infants themselves: what capacities would they need to allow various adults to handle, feed, and care for them?

Hrdy begins by noting that human infants and ape infants share some morally important emotions – but ape infants lose them while in human infants and children they continue to develop.

Human babies seek out faces, for example, and will continue to gaze into the eyes of another person for extended periods, sometimes mimicking facial expressions. Primatologists have observed mutual gazing in monkeys and other apes and even one monkey following the gaze of another (51). But while infant chimps (57) and even some monkeys (58) will mimic facial expressions of human experimenters, they typically lose interest in doing so (57). Human infants, on the other hand, get better and better at reading and making facial expressions, increasingly making facial expressions that will illicit interest and approval from others.

As widely evidenced in the primate literature, chimps and bonobos will share food but almost always by allowing another to take some of the divisible food items one is already eating or preparing to eat. Young children will impulsively proffer food to others, and sharing is increased by face-to-face contact. Like tamarins and marmosets, other primate species in which alloparenting exists, human children are interested not just in what they have received but also in what others have received (96–97). Tamarins and marmosets are much more likely than chimps, for example, to pull food trays towards adjacent cages that are occupied by a fellow conspecific, whether they are related to this individual or it is a stranger.

It is Hrdy's contention that such prosocial emotions, together with other factors in the early *Homo* environment, led to the development of cooperative breeding and alloparenting in the *Homo* genus and that this cooperative breeding co-evolved with the development of the much deeper and richer set of prosocial emotions that we observe in humans and to a lesser degree in tamarins and marmosets.

The form of alloparenting that emerged in the hominin line was, Hardy claims, due to a particular set of environmental factors. The first several factors are the likely small population of early *Homo* species with limited generic diversity, ranging over a large territory that would have made intergroup hostility needlessly costly. Added to these factors would have been the relative scarcity of easily digested soft foods in environments that required a good deal of walking to locate and gather such food. These factors would have been further exacerbated by unpredictable rainfall and seasonally fluctuating food availability, as well as by the fact that fibrous and difficult to digest tubers probably formed an important part of the diets of early hominins. If we assume in these conditions high infant and childhood mortality, we can also assume that cooperative breeding would have enabled much faster breeding and hence much higher reproductive rates for reproducing females.

Combining with these environmental factors would have been long lifespans for those hominins who survived childhood, together with low turnovers in group membership. This would have made available within highly stable groups grandmothers and great-aunts no longer able to bear young of their own but with interests in other infants and children within their group. Slowly maturing infants and children would have themselves provided a project requiring long-term commitments from more than just the mother, as is possible in other hominid species, thus giving hominin mothers an interest in sharing their young with other group members. In other hominid species, other females are often interested in the infants of mothers, but mothers are not interested in letting go of their infants. In alloparenting, mothers give up contact with their infants for extended periods of time and other adults give these infants extended amounts of care.

Hrdy's view also assumes hominid intelligence and selfishness were bounded by prosocial emotions that increasingly responded to an environment not particularly rich in one-off prisoner's dilemma situations but instead quite rich in continued commitments to the shared projects of survival, reproduction, and alloparenting. On Hrdy's view, humans did not evolve to solve one-off prisoner's dilemmas. What our evolutionary past has left us with instead is a complex matrix of functionally related moral emotions geared towards long-term commitments to ongoing projects in relationships with others, the chief project being the rearing of children.

The aforementioned environmental factors gave hominin species an increased evolutionary pay-off for the more fully developed prosocial emotions we find in *Homo sapiens* and, in particular, their mutually reinforcing nature as a matrix of prosocial emotions. The most salient emotion in Hrdy's discussion is trust: mothers trusting other group members with their children and children trusting other group members to feed and care for them. This kind of trust requires more highly developed capacities for reading the intentions of others, as well as more highly developed forms of empathy that would have enabled humans to both understand and appropriately respond to the intentions, needs, and desires of others and to know what others expect of us and to know when we (or they) have failed in what is expected. Tied to this would have been a capacity for being concerned about the approval or disapproval of others, as well as a concern with the interests of others and what others get relative to us and relative to each other. Important for infants, mothers, and thus alloparents more generally would have been the sense of security and joint commitment that emerges as others and we ourselves do what we are all expected to do.

Although Hrdy talks about mutual expectations and commitments, as well as concern with the approval and disapproval of others, she does not specifically talk about a sense of obligation or the felt externality of moral reasons. It is her contention that the morally rich set of emotions that she discusses emerged before language and hence before the articulation of moral rules. If Hrdy is right about the rich matrix of human moral emotions and the relative unimportance of one-off prisoner's dilemmas in the early *Homo* environment, human morality is unlikely to reduce to some version of contractarianism or even to some version of contractualism, forms of moral naturalism we will examine in the next chapter. There is much more content to our human moral emotions than simple respect for other humans as our rational or moral equals. Instead, it is easier to see such respect emerging out of empathetic caring, trust, and a concern for fairness. By the same token, it is easier to see human morality itself arising out of all these same emotions, plus rationality, plus human language and moral argument. Enlightened self-interest and rational contracts might reinforce and shape morality, but they are not likely to be at its foundation, if that foundation is as morally rich as Hrdy's cooperative breeding hypothesis supposes it to be.

Here we should distinguish between the evolutionary or historical foundation of human morality and its justificatory foundation. Conceptually, these are two very different things. An important part of what we are doing over the course of this book is building the idea that for EMR, the biological explanation for morality in general is linked to the justification of human moral codes and judgments. Hence, our overarching concern with moving from is to ought, a concern addressed most fully in our final chapter.

Wild justice

Closer to EMR than either de Waal or Hrdy are Bekoff and Pierce (2009). Like us, Bekoff and Pierce think animals are capable of morality, and that there are moral natural kinds. For them, however, the natural kinds are particular patterns of behaviour caused by particular kinds of psychological capacities caused by particular kinds of neurochemical events in the brains of the animals that manifest the behavioural patterns.

Bekoff and Pierce begin with the observation that numerous animal species, including humans, appear to share capacities for cooperation, empathy, and justice. They then argue that these capacities, in intelligent and social species, lead to the same general patterns of behaviour with the same kind of general functional outcome, namely group cohesion that successfully mediates between the competing goods of individual group members and the common good of the group as a whole. Along with Churchland (2011), they note that these capacities often seem to be the result of the same kinds of neurochemical events in the brains of the animal species at issue. Their conclusion is that if this set of capacities in humans amounts to a capacity for morality, other animals share in this same general capacity. This means that if morality exists for humans, it likely exists for other animals as well, those who like us are intelligent, social, and live in enduring cooperative groups.

Like de Waal, Bekoff and Pierce premise their argument for similar capacities on evolutionary continuity, but they move from similarity to sameness of kind based on a further premise: if psychological capacities arise from the same sorts of neurophysiological causes and have the same functional effects, they must be the same general kind of capacities. Animals thus do not possess capacities that are similar to (or the building blocks of) the moral capacities in humans but the very same kinds of moral capacities.

Near the beginning of their book, Bekoff and Pierce provide a working definition of morality:

> We define morality as a suite of interrelated other-regarding behaviors that cultivate and regulate complex interactions within social groups. These

> behaviors relate to well-being and harm, and norms of right and wrong attach to many of them.
>
> (7)

By right and wrong, they mean behaviours that are expected in the sense that shattered expectations lead to shattered social consequences, at least in the short run. As for de Waal, conciliation is an important aspect of moral behaviour, except of course that de Waal would not think of such behaviours as being explicitly moral if they are not performed as part of following a human moral code. So, for example, one dog might bow to another when in the course of play behaviour it bits the other dog too hard:

> For instance, one way we know that animals have social expectations is that they show surprise when things don't go "right" during play, and only further communication keeps play going. For example, during play when one dog becomes too assertive, too aggressive, or tries to mate, the other dog may cock her head from side to side and squint. . . . [T]he violation of trust stops play, and play only continues if the playmate "apologizes" by indicating through gestures such as a play bow his intentions to keep playing.
>
> (121)

For Bekoff and Pierce, this counts as moral behaviour, whereas for de Waal and many other comparative biologists it would count as merely as prosocial behaviour.

A key claim of Bekoff and Pierce's position is that most comparative biologists are, like de Waal, drawing the line between prosocial behaviour and moral behaviours in the wrong place. Moral behaviours are, for Bekoff and Pierce, a subset of prosocial behaviours that can be defined as follows:

> within this huge repertoire of prosocial behaviors, particular patterns of behavior seem to constitute a kind of animal morality. Mammals living in tight social groups appear to live according to codes of conduct, including both prohibitions against certain kinds of behavior and expectations for other kinds of behavior. They live by a set of rules that fosters a relatively harmonious and peaceful coexistence.
>
> (5)

Animals capable of morality, say Bekoff and Pierce, "form and maintain complex networks of relationships, and live by rules of conduct that maintain a delicate balance, a finely tuned homeostasis" (3).

Bekoff and Pierce have three kinds of behavioural patterns in mind, each of which they explore at some length in their book. For us the first cluster, the cooperation cluster, is the most problematic, because cooperation is not always directed towards moral goods. Killing prey is not morally good, nor is suppressing a large social underclass, even with a high degree of social cooperation among the oppressors. In any case, it is the adaptive traits that make possible the suite of cooperative behaviours (grooming, hunting, etc.) that Bekoff and Pierce are most interested in: "honesty, trust, punishment and revenge, spite and the negotiation of conflicts" (59). Again, punishment may be important for reinforcing moral behaviour, but on our view it is not clearly moral in and of itself. And we should again note that honesty and trust can be used, among an oppressing class, as important tools for maintaining social oppression, which we do not think should count as a moral good.

An interesting question for Bekoff and Pierce is why not: systems of oppression can be highly stable and highly cooperative. While offensive to the moral sense of the oppressed class, such finely tuned networks of behaviour may not be offensive to the moral sense of the oppressing class. If our moral sense ultimately defines what is morally right and wrong, what are we to say when our moral sense gives us this sort of deeply divided output? One could argue here, of course, that cooperative oppressive behaviour would be even more finely tuned were it not oppressive; but again, it is not clear that this is so, and even if it is, what gives independent (moral?) value to being more fine-tuned as opposed to less fine-tuned when it comes to networks of social behaviours?

Bekoff and Pierce's second cluster of moral behaviours is more in line with some of the natural moral values we discussed in Chapter 3. The "empathy cluster includes sympathy, compassion, caring, helping, grieving, and consoling" (87). This list seems to mix behaviours with psychological capacities, but Bekoff and Pierce see the two as being closely linked in terms of morality: moral behaviours require moral psychological capacities. Helping behaviour thus requires emotional and cognitive states on the part of the performer of that behaviour that are positively directed towards the well-being of the recipient of the behaviour.

The third cluster of behaviours, which is for Bekoff and Pierce a crucial aspect of morality, are behaviours involving justice. These are the behaviours that give their book its title, *Wild Justice*:

> Justice is a set of expectations about what one deserves and how one ought to be treated. . . . Our justice cluster comprises several behaviors related to fairness, including a desire for equity and a desire for and capacity to share reciprocally. The cluster also includes various behavioral reactions to injustice, including retribution, indignation, and forgiveness, as well as reactions to justice such as pleasure, gratitude and trust.
>
> (113)

Again, this is not a particularly well-defined list of behaviours, ranging again over both capacities and behaviours and including general feelings like (the natural good of) pleasure and general emotional responses to things that do not please us such as displeasure or indignation. When others cheat we do not like it, and sometimes we become indignant: but so too when our non-moral expectations are not met. In any case, Bekoff and Pierce offer a good review of some of the rich empirical work in this general area, and it is clear enough in their book what they mean to include in their justice cluster of moral behaviours. A good example, in addition to those involving fair play in dogs, is the capuchin grape experiment.

For Bekoff and Pierce, all three of these clusters are moral clusters because they involve similar behaviour patterns aimed at a similar end and based on similar neurophysiological causes. The end, again, is a finely tuned social homeostasis, which one might be tempted to say must therefore be, on this approach to morality, the summum bonum of all morality. An immediate problem with this is that homeostasis is a widespread biological and physical phenomenon, and in and of itself, there is nothing particularly moral about a biological or physical system in such a state. In arguing that animal morality is the same kind of thing as human morality, Bekoff and Pierce ignore the deeper theoretical question of what morality itself might be, apart from certain mammalian behaviour patterns that exist in some sort of a vaguely defined homeostatic balance with one another. Animals and humans both may be capable of morality, but what, exactly, defines morality itself and, in particular, moral goodness?

Although they do not say a lot about it, the homeostasis patterns that Bekoff and Pierce have in mind are ones that balance individual competition and individual needs against the needs or goods of other individuals in a social group, along with what might be thought to be the common goods of the group itself. But Bekoff and Pierce are vague on how this balancing is supposed to work within and across species, and when and how, exactly, it is supposed to be moral. For example, in studying robbing and dodging behaviour in rats, Bell and Pellis (2011) looked at experiments where two rats are in a cage and one of them discovers a small food item surreptitiously placed in the cage by the experimenters. The other rat, upon discovery of the first rat contentedly munching on the food item, will attempt to swoop in and steal the item away, behaviour which will then apparently cause the first rat to dodge the robbing effort of the second rat. Bell and Pellis argue that the situation is more complex, with the first rat continuously attempting to maintain a given distance from the second while the second continuously attempts to upset the homeostasis the first is actively seeking to maintain. The behaviour of each is affecting the behaviour of the other in a finely tuned balance of robbing and dodging. Sometimes robbing works, but it is generally better to be a dodger than a robber.

Robbing and dodging behaviour may prevent escalation to aggression, and it may best serve the needs of individual rats that sometimes come across surprising food items. But we might wonder if it is moral behaviour; it is social, it may be homeostatic, but it does not seem to be particularly moral. Bekoff and Pierce are focused on behaviour patterns with no explicit reference to what features in the environment these patterns of behaviour might arise in response to. Having your cheese and eating it too may be good, along with the occasional successful robbery, but it is not clear how or why either of these is a moral good. If the possibility of helping others, on the other hand, arises in a species' environment, it may be an independently good thing whether the members of the species notice it or not. Bekoff and Pierce might argue in the robbing and dodging case that there is no expectation of sharing, so there is no moral behaviour pattern. But the deeper question is still what triggers such expectations in some circumstances but not others – how and why do some homeostatic patterns cross over into morality, while others do not? As we saw in Chapter 3, if a captive rat is struggling, the rat presented with the food item typically frees the first rat before both become interested in the food. What makes not sharing in the robbing and dodging experiment morally good, or at least morally permissible, while failing to free your struggling fellow species member seems to be morally bad in the captive rat experiment? In each case, the rats may have psychological capacities that tell them what to do, capacities that are linked to social homeostases with joint expectations about how things are to normally play out in these sorts of homeostasis situations. But why are some psychological capacities and expectations "moral" while others are not? Bekoff and Pierce seem to need a homeostatic summum bonum at the bottom of their approach to morality that is hard to define and that would moreover lead their approach towards a foundation of moral truth that goes beyond whatever particular capacities might tell us, whatever our particular species happens to be, about how we morally ought to behave.

So while Bekoff and Pierce's moral sense approach to moral naturalism bears some important similarities to EMR, there are important differences. Like EMR, Bekoff and Pierce suppose that other animals, besides humans, are capable of morality. They also tie morality to natural kinds and, in particular, to clusters of related natural kinds. But whether we begin an evolutionary approach to morality with clusters of natural moral capacities or clusters of natural moral values makes a difference to both the explanatory and justificatory aspects of an evolutionary and thus a naturalistic approach to ethics.

On the explanatory side of such an approach, on Bekoff and Pierce's theory there are no natural moral values, just natural capacities tied to patterns of behaviour that appear to be similar to one another. Behavioural patterns develop, these patterns have a positive effect on inclusive fitness for the species in question, and there are even greater gains in inclusive fitness to be had if these behaviour patterns can be made more plastic through the evolutionary development of more complicated psychological mechanisms for generating them. Insofar as these psychological developments might lead, in the human species, to a belief in moral values, this belief must be accounted for by some version of the error theory, unless we suppose that moral capacities, whether they know it or not, are aimed at a finely tuned homeostasis yet to be defined by evolutionary biology.

For EMR, moral capacities develop in environments containing moral values. These moral values will be part of the explanation for why the capacities develop the forms and structures that they do. Just as a species' physical terrain will be an important part of the explanation for why its legs have the structure that they do, a species' moral terrain will be an important part of why its moral capacities have the forms and structures that they do. As we will argue in the next chapter on reductionist approaches to morality, it is unlikely that morality is ultimately reducible to behaviour patterns. Moreover, we greatly increase the explanatory potential of an evolutionary approach to ethics if we assume that moral capacities are developing in response to environmental pressures that are not simply caused by the capacities themselves or by the patterns of behaviour that we might suppose the capacities arise out of and begin exerting an influence on.

Although Bekoff and Pierce think that some other animals have moral capacities of the same general kind or kinds as ours, they also think that moral capacities are species-specific:

> We advocate a species-relative view of morality. Each species in which moral behavior has evolved has its unique behavioral repertoire. The same basic behavioral capacities will be present – empathy, altruism, cooperation, and perhaps a sense of fairness – but will manifest as different social norms and different behaviors (e.g., different grooming patterns of unique ways of expressing empathy). Despite some shared evolutionary history, wolf morality is different from human morality and also from elephant morality and chimpanzee morality.
>
> (Bekoff and Pierce, 2009, 19.)

On the one hand, say Bekoff and Pierce, current research "shows that human moral behavior is much more 'animal-like' than our common-sense assumptions would suggest" (31). On the other hand, in current research "[i]n other areas of comparative biology (e.g., auditory and olfactory communication), the human-as-gold-standard has proven deficient because each species has its own distinctive capacities adapted to its own particular environmental and social circumstances" (19–21). Just because other animals do not possess exactly the same forms of emotional intelligence that we humans do, this does not mean that they cannot be more emotionally intelligent than us in ways that are unique to their species, and so too with morality. Precisely because animals are not empathetic in exactly the same ways that we humans are, they may exceed us in other ways when it comes to the more general phenomenon of empathy. Clever Hans, for example, might be much better at reading human facial expressions or related stress levels than we humans are ourselves.

These points raise questions about what counts as evolutionary continuity and what counts as evolutionary sameness when it comes to a general capacity or set of capacities for morality. While there are arguments available to Bekoff and Pierce for drawing the lines between similarity and sameness as they do, the arguments available to EMR are more simple and straightforward: there is a general level of sameness across different moral capacities in different species because these capacities are all responding to the same kinds of things – natural moral values – that can be found in similar kinds of environmental and social circumstances that exist across a wide variety of species that are social and intelligent.

While we think Bekoff and Pierce's approach to wild justice is an interesting theoretical move in the right direction, we think that they do not go nearly far enough with their proposals. This move to a deeper level of explanation enables us to begin to think about when and how moral values like helping another might have first appeared, and how in shifting environmental and social circumstances that value might have itself developed to become more complex as the species' adaptations to it became more complex themselves, with these levels of complexity sometimes creating spaces for new kinds of moral values that might complement and extend the reach of the earlier ones. Thinking about moral values as natural moral kinds that arise in the environmental and social circumstances of species that are social and intelligent may enable us to better predict how, where, and why new moral behaviours and capacities develop, predictions unavailable to us if we begin from behaviour patterns alone. At this point in our development of EMR, we can only raise these possibilities – our argument in this book is that EMR is worth considering as the basis of a research program.

On a related point, we think that EMR's approach to natural moral goods also promises to provide a better account of the kinds of homeostases that are no doubt part of morality. Morality does centrally concern itself with adjudicating between competing individual interests and the interests of the group as a whole, at least for species that are both social and intelligent. Natural moral values and the responses to them are a necessary part of what pushes social and intelligent species towards moral equilibria that are appropriate for their particular environmental and social circumstances. Natural moral values do not define what this kind of equilibrium should be, but they do enable the organisms involved to get as close as they can to appropriate equilibrium points for their particular circumstances. For humans, this means talking and arguing about such values, a point we will further develop in the remainder of this book. When we know more about such equilibrium points and the processes that go into reaching them, we may be in a better position to make general claims about their nature. They may thus emerge as a kind of natural moral good themselves, available to social and intelligent organisms capable of evolutionary biology for their moral guidance.

This brings us to our last point in this section. If what we hope to account for is morality itself, then Bekoff and Pierce's approach, grounded as it is solely in behaviours and capacities, raises significant problems for moral justification. These are the same problems that face the standard view more generally. If our moral beliefs are ultimately grounded in our moral capacity, our moral beliefs are only true because we have come to think they are true, not because they in fact are. This makes it hard to account for historical changes in human moral thought as genuinely progressive or regressive, and it threatens an evolutionary approach to morality with the dilemma of either falling prey to the naturalistic fallacy or adopting some version of the error theory as the only way to avoid the fallacy. We take up the problem of accounting for moral progress briefly in the final section of this chapter and at greater length in the chapters that follow. We take up the problem of the naturalistic fallacy in the final chapter of the book.

Moral sentiments and moral progress

Nichols (2004) provides a good example of an experimental approach to moral philosophy that follows de Waal in making morality derivative from moral instincts and emotions or, as the history of moral philosophy would refer to these aspects of human nature, the moral sentiments. In attempting to provide empirical evidence in favour of particular structures for these sentiments, Nichols' view may provide a missing link between de Waal's pillars of morality and the moral systems of human societies that result from language, thought, argument, and social agreement. Although EMR is committed to natural moral values that affect the appearance and development of moral instincts and emotions, it can remain neutral on the details of Nichols' account of the moral sentiments and what he calls the sentimental rules of morality. We include a discussion of Nichols in this chapter because it is an influential view that reduces the content and motivational force of moral rules to moral instincts and emotions, in contradistinction to EMR. There is more psychologically detailed work in this area, such as Hauser (2006), but again, the details of this work do not much matter for our purposes here.

Nichols begins with the point that from quite a young age, children distinguish between moral rules and conventional rules. They do this before they are able to take the perspective of others, a key aspect of empathy according to some philosophers. Because empathy might be supposed to be the key emotion for morality, this seems problematic. But as we have seen in earlier sections of this book, evolutionary psychologists like de Waal (2009) argue that there are phylogenetically earlier forms of empathy that do not involve perspective taking. Nichols himself continues on to discuss a variant of this idea from Blair (1995). But Nichols fails to note that even very young children may have noticed that the adults around them get particularly upset when some rules but not others are violated. Not surprising, the more interesting and important rules will typically involve such moral concerns as not harming others.

Following Blair (1995), Nichols hypothesizes that humans have an instinctual response against harming others, and, as we have seen in earlier sections, something like this hypothesis seems quite likely to be true. For Nichols, the cognitive component of this instinctual response supplies the basic content to a normative moral theory that humans acquire as our minds develop. This moral theory is structured around what Nichols calls core moral judgments, like the judgment that hitting the child next to you is wrong in a way that chewing gum in class is not wrong. In Stanford's terms, children as young as 2 take moral reasons to be externally binding on them in a way that conventional reasons prohibiting actions are not. The obligatory force of moral reasons cannot simply be undone by someone in a position of authority declaring that such reasons no longer apply. If a teacher says chewing gum in class is alright, it is alright. But it is not alright to hit the child sitting next to you, even if the teacher says that it is.

Core moral judgments, according to Nichols, are "sentimental rules," rules with moral content that line up with strong human emotions in their favour. Both the content of the rules and the emotions that make the rules motivationally effective are central parts of human nature, and they are tightly linked from a very early age. Together, they provide an account according to which "morality derives from the sentiments" (29).

Central to sentimental rules is what Nichols calls the concern mechanism: the capacity to detect and be moved by the suffering of others (41–46). According to Nichols, this mechanism does not require taking the perspective of the other, but it does require the capacity to attend to the negative mental states of others. The argument is that in responding to the distress of others and helping them, small children are not simply responding to emotional contagion or their own distress.

Nichols gives some interesting empirical evidence for these developmental claims from Zahn-Waxler et al. (1992). Between the ages of 13 and 25 months, children increasingly respond to the distress of others with sad facial expressions and sympathetic responses, followed by an increasing correlation at 18–25 months old between such responses and efforts to comfort the individual who appears distressed. This is similar to the behaviour of Yoni already discussed here and in de Waal (2009). Although Nichols does not go so far as to claim this, if the concern mechanism is the source of morality, this mechanism would seem to exist across a number of species. If it is to supply the content of our most basic moral rules and their effective motivational force, it seems plausible that other species might also exhibit some degree of moral agency on a view like the one developed by Nichols.

As part of testing his theory of the moral sentiments, Nichols makes the claim that if the concern mechanism is at the core of morality, moral rules that violate it are less likely to be historically stable than rules that are consistent with it. Although we think something like this is likely to be true, rules that cause human suffering are also likely to threaten social cohesion, while rules that respond to human suffering are likely to enhance social cohesion. The effect that Nichols seeks to explain with his sentimental theory of morality is likely to be overdetermined, something which he himself notes in passing later in his book (155).

Nichols' main empirical claim, largely unargued, is that moral rules historically progress to become both more inclusive (e.g., Western thought has come to include animals as having moral standing) and less violent (e.g., forms of punishment have become less physically injurious). His claim is that it is hard to explain these changes without invoking the concern mechanism as a deeply embedded part of human nature.

Other possibilities he considers are two forms of moral realism. Version 1 of moral realism claims there are moral facts and that historically humans have gotten better at detecting these facts. The main problem with this view, he says, is that no one knows what a moral fact might look like. EMR provides a direct response to this problem. If EMR turns out to be correct, moral facts were in front of us from before the time we acquired language, and if they have disappeared from sight over the course of Western philosophy, this is because Western philosophy has been looking for them in the wrong places. The related practical problem is that highly individualistic Western societies have also increasingly lost sight of the natural moral values that ought to matter to biological creatures like us. Because EMR seeks to include within its ambit views like those of Nichols and de Waal, like these views it also needs to be able to account for the fact that moral rules often go astray in ways that make later progress possible. But just like these views, EMR can point to other aspects of human nature, such as selfishness, nepotism, and concerns about social hierarchies and social standing, all of which may also be expected to strongly influence social rules. On any naturalistic approach to morality, there will be plenty of ways that moral rules can become morally corrupt.

The second form of moral realism Nichols considers, Version 2, comes from Railton (1986). According to Railton, the facts underlying moral truth are facts about human rationality and the fact that such rationality has a social bias. Certain moral rules, as a matter of fact, can be justified rationally from a socially impartial point of view, while others cannot (Nichols 2004, 161–162). Nichols has several arguments against this view, but none seem to us conclusive and they do not affect our argument here. In accord with the contractualist theories of morality we consider in the next chapter, we are willing to accept that humans are concerned, at least some of the time, with impartial reasons and that such reasons can lead to historical changes in moral rules in more impartial directions, such as concern for animals or for less violent forms of punishment.

The main point here is that for Nichols, there can be moral progress, but any such progress cannot be tracking moral truth because the ultimate truth about morality is that it is derivative from human moral capacities such as the concern mechanism. That such a mechanism is a part of human nature is empirically true, but it is not a moral truth. For Nichols we do not need to hypothesize anything like moral realism to get human morality, but we do need to hypothesize something like the concern mechanism.

If EMR is true, we would also expect to see historical effects in terms of changes to moral rules over time. These might sometimes be explained simply by our moral instincts, but the fuller explanations are much more likely to be multifaceted. In particular, ongoing efforts to bring our moral and empirical judgments into wide reflective equilibrium will bring with them a variety of changes in our moral rules, some of them progressive and some of them not. Language, thought, and argument create a rich overlay of moral distinctions, from the highly specific to the highly general, extending well beyond the moral instincts upon which these distinctions are built. Such widely reflective considerations are largely absent from Nichols' book. We will turn to such considerations in the last chapter of this book, but we will argue in the meantime and in the next several chapters that natural moral values might sometimes play a role in episodes of social change. The main difficulty for our argument in the next chapters is to tease out a possible role for natural moral values in social changes apart from moral emotions, social perspective taking, and moral arguments aimed at establishing greater consistency in our judgments in wide reflective equilibrium. In any interesting social change, all these factors are likely to play a role.

Having raised the issue of moral progress, we need to consider briefly Shermer's *The Moral Arc* (2015). This book offers a more historically detailed version of the sentimental argument for moral progress than the one to be found in Nichols (2004), but it does so in a way that might seem to involve a form of EMR. We do not want our argument here to be confused with Shermer's, which we take to be highly problematic and somewhat confused.

Unlike Nichols, Shermer plays fairly fast and loose with the distinction between is and ought. Early in the book (12–13), he offers a general moral principle that what is fundamentally morally good and right is that which improves the flourishing of sentient beings as individuals. He includes within what is good for the individual well-functioning social relationships, because individuals are linked to other individuals through kin selection and reciprocal altruism (39). The fundamental moral principle then functions as a key premise in rational arguments that lead to moral progress:

> *Descriptively*, science tells us that human beings have an evolved, innate drive to survive and flourish, and that one of the most necessary and primal requirements among the many preconditions for life, health and happiness for most people is a loving bond with another human being. *Prescriptively*, we can say that granting only a select group of privileged people the right to fulfill this evolved need – while simultaneously depriving others of the same basic right – is immoral because it robs them of the opportunity to fulfill their essence as evolved human beings.
>
> (33)

Shermer does not explicitly cite his principle in the course of this argument for gay rights, but it is clearly there, as an unstated but necessary premise: robbing some sentient individuals of their right to flourish is wrong precisely because (fundamental principle of morality) it is morally right and good to improve the flourishing of all sentient individuals. This missing premise is what gets us from his descriptive "is" to his prescriptive "ought."

Shermer's claim is that we can come to know that this principle is true through scientific methods. What he has in mind is that the principle is a "law" of human nature (121–122 and 205–207) that can be discovered biologically and psychologically. Psychologically, humans in fact care about the flourishing of other sentient creatures as individuals, and biologically humans are, as a matter of fact, built this way to enable our survival and flourishing as individual members of our species. How this empirical law of human nature (if it is indeed one) becomes a moral law is not made clear by Shermer. What he seems to have in mind is something like Nichols' argument: given our moral emotions and given our capacity for language and reason, we are built to think in accord with what Shermer identifies as the fundamental principle of human morality. Once we discover this fact about ourselves, we are then free to discover empirically what sorts of social arrangements are best in accord with the principle that governs our biological natures as human beings, just as in the preceding quoted argument about equal marriage rights for gays and lesbians. What gives the principle behind this argument its moral force comes from within, so to speak: we ourselves give it moral force because we ourselves are constructed to give it moral force. In short, things are morally right or wrong only because as we human beings are constructed to think them so. Morality derives from the human moral sentiments, lawful or otherwise.

Once this confusion is unpacked, Shermer's argument has more in common with Nichols's argument than with ours. For EMR, natural moral values exist across species, and they are responsible for traits like the ones Nichols and Shermer appeal to in their approaches to human morality. But like Shermer's approach, EMR also faces the problem of getting from descriptive facts about the biological world to morally prescriptive oughts. It is to this problem that we now turn.

Bibliography

Bekoff, Marc, and Jessica Pierce. 2009. *Wild Justice: The Moral Lives of Animals*. Chicago: University of Chicago Press.

Bell, Heather, and Sergio M. Pellis. 2011. "A Cybernetic Perspective on Food Protection in Rats: Simple Rules Can Generate Complex and Adaptable Behaviour." *Animal Behaviour* 82 (4):659–666.

Blair, R. 1995. "A Cognitive Developmental Approach to Morality: Investigating the Psychopath." *Cognition* 57:1–29.

Churchland, Patricia. 2011. *Braintrust: What Neuroscience Tells Us about Morality*. Princeton: Princeton University Press.

de Waal, Frans B.M. 1996. *Good Natured: The Origins of Right and Wrong in Humans and Other Animals*. Cambridge, MA: Harvard University Press.

de Waal, Frans B.M. 2006. *Primates and Philosophers: How Morality Evolved*. Princeton: Princeton University Press.

de Waal, Frans B.M. 2009. *The Age of Empathy: Nature's Lessons for a Kinder Society*. New York: Random House.

de Waal, Frans B.M. 2016. *Are We Smart Enough to Know How Smart Other Animals Are?* New York: Norton.

de Waal, Frans B.M., and Pier Francesco Ferrari, eds. 2012. *The Primate Mind: Built to Connect with Other Minds*. Cambridge, MA: Harvard University Press.

Flack, Jessica, and Frans B.M. de Waal. 2000. "'Any Animal Whatever': Darwinian Building Blocks of Morality in Monkeys and Apes." *Journal of Consciousness Studies* 7 (1–2):1–29.

Hauser, Marc D. 2006. *Moral Minds: How Nature Designed Our Universal Sense of Right and Wrong*. New York: HarperCollins.

Hrdy, Sarah Blaffer. 2009. *Mothers and Others: The Evolutionary Origins of Mutual Understanding*. Cambridge, MA: Harvard University Press.

Joyce, Richard. 2006. *The Evolution of Morality*. Edited by Kim Sterelny and Robert A. Wilson, *Life and Mind: Philosophical Issues in Biology and Psychology*. Cambridge, MA: The MIT Press.

Langergraber, Kevin E., David P. Watts, Linda Vigilant, and John C. Mitani. 2017. "Group Augmentation, Collective Action, and Territorial Boundary Patrols by Male Chimpanzees." *PNAS* 114 (28): 7337–7342.

Nichols, Shaun. 2004. *Sentimental Rules: On the Natural Foundations of Moral Judgment*. Oxford: Oxford University Press.

Railton, Peter. 1986. "Moral Realism." *Philosophical Review* 95:163–207.

Schmelz, Martin, Sebastion Grueneisen, Alihan Kabalak, Jurgen Jost, and Michael Tomasello. 2017. "Chimpanzees Return Favors at Personal Cost." *PNAS* 114 (28): 7462–7467.

Shermer, Michael. 2015. *The Moral Arc: How Science and Reason Lead Humanity to Truth, Justice, and Freedom*. New York: Henry Holt and Company.

Stanford, P. Kyle. 2018. "The Difference between Ice Cream and Nazis: Moral Externalization and the Evolution of Human Cooperation." *Behavioral and Brain Sciences* 41. https://doi.org/10.1017/S0140525X17001911.

Stingl, Michael. 2016. "Smartening Up: Animal Cognition from an Evolutionary Point of View." *The Quarterly Review of Biology* 91 (4):487–490.

Zahn-Waxler, C., M. Radke-Yarrow, E. Wagner, and M. Chapman. 1992. "Development of Concern for Others." *Developmental Psychology* 28:126–136.

5 Reason, rational contracts, and selfish genes

Human morality

Human beings are rationally self-interested as well as more generally rational. We expect to find causes behind events and reasons behind actions, and we expect such causes and reasons to exhibit various forms of consistency. Moral reasoning is no exception. But while we typically demand moral reasons from one another when we are in morally challenging circumstances, the moral force of such reasons may stem from more than one source. As we are arguing in this book, a biologically based capacity for recognizing naturally occurring moral values may be a fundamentally important source of our human interest in and capacity for moral reasoning. There may be other sources as well, some of which we will explore in this chapter.

From the point of view of EMR, Western philosophy suffers from a peculiar sort of mental cramp when it comes to moral values. Common sense thinking would suggest to us that moral values are independently real aspects of the world we live in as humans. But according to Western philosophy, this is a metaphysical mistake. Moral values cannot be as simple as the fact that helping another is a good thing. If they are anything at all, moral values have to be something more sophisticated, some sort of artefact of the higher human capacities of thought and language, or something that is only accessible through thought and language. EMR can partially agree with such a view; the further along a moral trajectory a species travels, the more complex the moral values will be that come to structure its social environment.

EMR takes tensions between group and individual interests to be a central problem of morality. With rationality comes the self-conscious possibility of separating one's own individual interests from the interests of others, and having separated them, the possibility of wondering which to favour over the other, should they come into conflict. Moral reasons and their force as reasons will significantly depend on the specifically human capacities that enable us to articulate, argue about, and ultimately agree on reasons for acting in certain sorts of ways and not others. In contrast, species without rationality will have no basis for thinking about moral values in this way, although they may act in conformity to them.

In this chapter, we consider several additional sources of human morality that we might suppose to have emerged from the evolutionary processes that produced modern humans. Approaches to morality based on these other sources of moral behaviour typically reduce morality to some uniquely human basis. Popular endpoints for such a reduction are human reason or human rationality. It is our claim that while these other sources of morality are important, they cannot by themselves completely account for morality. Morally significant phenomena appeared in the biological world before humans did. On the other hand, something of significant moral consequence did occur with the advent of human reason and the moral values of autonomy and moral responsibility. Still, there is no reason to suppose that the human species is at the end of its moral trajectory. Moral values and capacities of greater complexity may be biologically possible: if not on our trajectory, then perhaps on another. From a biological point of view, human beings are neither the beginning nor the endpoint of morality.

Human reason and moral truth

We encountered one important kind of reductionist approach to morality at the end of Chapter 3. For Peter Singer, human reason is at the core of impartiality and thus of morality. This means that in our current biological world, morality can only be human morality. There may of course be aspects of reason itself that are not accessible to humans, and so morality itself may extend beyond human morality. Yet insofar as humans are reasonable, they are moral animals. Although other animals might be objects of moral concern, they can only be limited subjects of such concern because their capacity for reason is limited. Moral truths are truths of reason, and so to recognize and be moved by moral truth one must be reasonable. If moral acts are the right acts done for the right reasons, animals cannot be moral agents. This will seem especially true if we tie moral agency to moral responsibility, and Singer's appeal to reason is one prominent way to do exactly this. Animals cannot reason, so they cannot be held responsible for their behaviour, so they cannot be moral agents. This makes morality as we know it an exclusively human phenomenon, except for the possibility that other species, on this planet or another, might evolve in such a way that their members also develop the capacity of (non)human reason.

This approach takes moral truths to be something like mathematical truths, a move Singer (2011) makes in his new afterword to the reissued *The Expanding Circle*. From an evolutionary perspective, this is problematic for the reason given in Joyce (2006, 182–183). It is relatively straightforward to see how numbers and other aspects of mathematics could be part of an environment in which an interest in numbers could have evolved (Singer 2011, 89). Four hunters go behind a bush and only three come out. Baboons can apparently count this high but no higher, a feature of the baboon mind that human hunters in larger groups can exploit to their advantage. Yet our evolved capacity for morality seems not to have evolved in this way, according to either Singer or Joyce. Unlike numbers, moral goods do not seem to have been the kinds of things that organisms might encounter in their environments and that might thus matter to them and to their successful reproduction. Singer responds to this point by arguing that moral truth is only accessible through pure reason, while Joyce argues that moral goodness is an illusion necessary to human interaction and survival.

Reason's relationship to a non-illusory form of moral goodness may of course have developed independently of encounters with environmentally present moral goods, something that Joyce fails to take into account by assuming that moral truths, if they exist, must be somehow reducible to other empirical truths. Joyce is criticized on this point by Shafer-Landau (2012), Brosnan (2011), Kahane (2011), and Wielenberg (2010). Reason may have evolved for some other purpose, or it may be a spandrel of some sort, one which gives humans the capacity to recognize moral goodness as a property that somehow supervenes on, or is otherwise correlated with, particular empirical properties of objects that we humans do regularly bump into in our environments and that do matter to our survival. The property of moral goodness, for example, might somehow supervene on natural properties like responding sympathetically to the pain of others without its being reducible to such properties.

Without a precise understanding of this supervenience relationship, something that has proved to be notoriously hard to provide, moral metaphysics and epistemology face the fundamentally sceptical challenge that Joyce (2006), Ruse (1986), Harman (1977), and Mackie (1977) have forcefully posed for it. It is more empirically likely that the human mind is creating and applying the moral properties of goodness and badness to things that we humans either really like or really do not like (where these likes and dislikes are linked to the deep effects of the natural properties of these things on our lives together) than that the things involved here actually possess extra, non-reducible, and supervenient properties that turn out to be apprehensible by reason alone.

On our view, both the metaphysical idea of the supervenience of moral properties on natural properties and the empirical response to it to be found in naturalistic views like Joyce's are mistaken. If EMR's form of moral naturalism is right, moral goods arise in the environments of certain kinds of social organisms. If we want to know what moral goodness is, we must begin with an empirical investigation of these kinds of good things, not with a semantic analysis of the predicate of moral goodness. Goodness is not fundamentally in our heads or in the land of abstract metaphysical truths. It is firmly ensconced in the real world as part of the evolution of certain kinds of cooperative networks of organisms. Certain kinds of cooperative environments produce moral values.

It is philosophically possible that moral values, as they are understood by contemporary metaphysical moral realists like Enoch (2011) and Shafer-Landau (2012), are nomological danglers or phenomena that do not fit into the system of established natural laws. If we suppose along with Nagel (1974) that mental states are nomological danglers only visible from a subjective point of view, we might suppose that moral values are nomological danglers that are only visible from a special kind of subjective point of view, what we might call, following Baier (1958), the moral point of view, or following Nagel (1986), the view from nowhere. Danglers do of course have to dangle from something, and this something will be part of the world of causes and effects. So Shafer-Landau and others are right to argue that it is not so easy to imagine that we could have exactly the same set of moral beliefs we currently have even if they were false, because if they were false, there might not be the right sorts of things in the causal world for moral values to dangle from, and so we wouldn't have the beliefs that we have.

On our view, moral values are not nomological danglers. For EMR, moral properties can precede the capacity to detect or recognize them, and they may thus have causal efficacy in a way that mental states would not, if they were nomological danglers. Being moved by the interests of others is already by itself a moral value. The ability to detect and respond to this value pushes species on a moral trajectory towards the development of a moral point of view. As natural kinds, moral values are fully part of the natural world of causes and effects, a point we will return to at the end of this chapter.

Whatever moral values might turn out to be, *moral reasons* are products of human reason. Once early humans were able to talk, they could say important things about their environment, including each other, as well as the moral values that significantly structured their relationships to one another: "You Tarzan," "Me Jane," "Tarzan not play fair."

This last statement makes a normative claim against Tarzan. What gives this normative claim its initial moral authority and makes it a *moral* reason is the negative moral value it gives immediate voice to, that of not playing fairly. Like "play," "fair" is a natural kind term. The denoting terms in "Tarzan not play fair" are thus all three directly referential. Play and fair play would have been important parts of our environment. We would have named them. Chances are good we would have been responding to them long before we were reasoning about them.

On our view, the moral authority of moral reasons is ultimately grounded in moral truth, and moral truth is ultimately grounded in natural moral values like fairness. What makes moral reasons *reasons* is our capacity to reason with and about the moral values in our environment. Reason brings with it the need for consistency among integrated sets of moral claims of varying levels of generality, what Rawls calls moral judgments in wide reflective equilibrium. Despite the apparent simplicity of our example referencing fair play, we do not believe that EMR is committed to some sort of one-to-one relationship of true moral claims to individual moral facts. At the level of individual moral judgments, moral truth may be no more interesting than propositional truth more generally: "Tarzan not play fair" is true if and only if Tarzan not play fair.

Regardless of how we are supposed to understand moral truth and moral reasoning, it seems clear that on some occasions the moral reason for acting in a particular sort of way may not be among an agent's reasons for doing what he or she would like to do. Tarzan may not want to play fair, and he may not care where this leaves Jane. Perhaps, following Singer, behind all such cases of bad moral judgment, we will find a failure of reason, and once this failure is pointed out to the agent in question, the agent, if he or she is reasonable, will concede that the moral reason in question ought to trump all of his or her other reasons for doing what he or she would otherwise want to do.

But there is a more immediate problem here from an evolutionary point of view. As Boehm (1999, 2012) argues, failing to take account of how your actions will affect others, in the context of human cooperative groups and human evolution, will typically pose a noticeable threat to group stability. Reason and philosophical argument aside, other group members will often have a joint interest in decreasing to zero the distance between the agent's reasons for behaving immorally and the moral reason for behaving as he or she ought to. Over many generations, this sort of social pressure may well lead to the formation of what Boehm calls the human conscience, and with it, an internalized moral push to do the right thing that it is hard for humans to resist. Although this process need hardly be assumed to be perfect, as indicated by the success of many sociopaths, we might follow Boehm in supposing that it will internalize most moral reasons for most people most of the time. But even if we agree with Boehm that the human conscience is an important part of human morality and the overall authority of moral reasons, we need not agree that morality itself is somehow reducible to beings like us having a conscience that demands that we do some things and not others.

Once human groups become much larger and more diverse than the small-scale groups suggested by our Tarzan and Jane example, moral reasons will continue to be articulated, developed, argued about, and agreed upon. People will generally use these reasons to guide their actions and to enforce the agreed upon moral norms defined by these reasons against those who would ignore them. Adherence to such practices will itself give an additional moral push to the moral reasons and moral norms that they involve. But as social relationships become more complex and different groups or social actors emerge who have more social and political power than other groups or actors, social norms will develop that favour, more or less subtly, some individuals or groups over others, for no good reason other than the fact that those who are more powerful will often have more sway over what moral norms are actually agreed to in larger and more diverse groups of human beings.

To the extent that we humans are reasonable and moral, these kinds of situations will lead to arguments over the consistency of our overall set of moral norms: do they really treat like cases alike, all things considered? The only way of ensuring that we have stripped all improper bias from our moral rules is by engaging in a critical process of wide reflective equilibrium, one which attempts to build the biggest possible set of consistent moral beliefs about the good and the bad and the right and the wrong. To the extent that we are reasonable moral beings, we will engage in this kind of critical process, with more or less success, given our social circumstances. To the extent that we are able to produce a set of considered moral judgments in a wide reflective equilibrium, these judgments will have as much authoritative moral force as is possible for humans at this point on our moral trajectory. That force comes, at its base, from the basic moral values that move us as moral creatures but also from the fact that the biggest possible set of consistent moral judgments we can achieve through a process of wide reflective equilibrium has the best chance of getting our moral situation right, relative to the basic moral values we started from. What pulls us away from such values is individual or group bias, and what pulls us back in the direction of basic values like fairness is the search for consistency among our overall set of moral beliefs.

To the extent that we can reach points of wide reflective equilibrium, we have the best reasons possible for thinking we are acting rightly in following our considered moral judgments. For humans at this point on our moral trajectory, this is as authoritative as moral reasons get. For EMR, it is also the closest we humans can get to moral truth.

The moral authority of conscience

Having begun this chapter with Singer's philosophical appeal to reason as the sole source of morality and moral truth, we turn in this section to what is perhaps the most philosophically minimalistic view of human morality and moral truth, the view that there are no moral truths, just moral illusions. This approach to morality is powerfully developed in an evolutionary context by Joyce (2006) and Boehm (1999, 2012).

As already noted, Boehm argues that the apparent authority of moral reasons comes from conscience. At the first appearance of modern humans, there was no conscience and thus no morality. Basing his argument on anthropological data from contemporary hunter-gatherer groups, Boehm argues that there was, however, individual autonomy, along with a desire to protect one's autonomy against incursions from the autonomy of others. Boehm is not supposing, as we do, that autonomy is itself a moral value: it is just something that matters greatly to those who possess it. What also mattered greatly to the earliest humans were highly nutritious food items, typically not easily obtained, and hence freely shared within the group as a whole once they were obtained.

Some individuals, typically men, were better than others at acquiring such food items, and based on their prowess in this regard, some of these men may have felt they were in a position to divide food less evenly than they otherwise might or to assert their autonomy over the autonomy of others in return for providing more equal access to such food items. When this sort of thing happens in groups of contemporary hunter-gatherers, the response of the group as a whole is swift, starting with ridicule, escalating to shaming, and eventually to shunning, banishment, or death. According to Boehm, these kinds of social pressures had evolutionary consequences, selecting for temperaments that readily responded to or did not engender in the first place these kinds of group punishments. Such temperaments were basically structured by guilt over abuses or potential abuses of one's autonomy over the autonomy of others. Out of such feelings of guilt developed our current human capacity of conscience.

With less empirical detail, Joyce (2006) offers a similar account of conscience and morality, significantly drawing out its implications for moral authority and moral truth. On such an approach to morality, there is no reason to believe that demands of conscience are either true or grounded in any authority that goes beyond the power of potential group punishments. The apparent moral authority of such demands is produced entirely within the human conscience, which turns the shame of group punishments into guilt. From the point of view of understanding morality as a natural phenomenon, this is the most minimalistic view of morality one could take: there really is no such thing as moral authority, only deeply internalized guilt and the concomitant appearance of an external form of moral authority.

For EMR, it is unlikely that humans evolved in a world devoid of moral values and more rudimentary capacities for recognizing these values. From this same point of view, conscience also seems like too slender a reed not to be easily bent by more self-interested motivations. Human conscience, all by itself, seems like a weak support for moral authority, even if we combine it with reason, as above, or with practical rationality, as we will see in the next section. Joyce may be right that if a story like Boehm's is true, we don't have to fear for our spoons: most of us value living together with others, and the cost of this is leaving our hands off the property of others. With strong enough consciences and other prosocial emotions, perhaps we need not fear that the practice of producing and responding to moral justifications would immediately collapse with the realization that there is nothing more to moral authority than psychologically internalized standards of social conduct.

But this approach to conscience is nonetheless wide open to Thrasymachus's challenge to morality in *The Republic* – if we accept Boehm's position, justice appears to be something imposed on the weak by the strong. Let conscience be your guide, but only until you are in a position to seize social power. At the level of hunter-gatherers, "the strong" is the community itself, imposing its will on any would-be big men. But if a would-be big man acquires the equivalent of Glaucon's Ring of Gyges – that is, some form of social power that enables its holder to impose his will on the rest of the community – Thrasymachus's challenge arises. Our spoons may be safe but not much else, depending on the whims of the tyrant who now rules over our lives. History tells us that this is not an idle worry. We may just have to swallow this worry as an unfriendly aspect of a natural world devoid of genuine moral values, but EMR gives us a more positive way to think about the worry and how we might respond to it as humans.

One might argue against EMR that on Boehm and Joyce's approach, we are not really lost without genuine moral values to guide us, because popular uprisings may eventually enable the many to reassert their combined power against the more concentrated power of the tyrannical few. When our individual autonomy is threatened, we will eventually band together to protect it, in a joint effort to return to society its prior semblance of equality. But this seems to miss the most popular aspect of popular uprisings: the fact that despotic power is manifestly unfair. Like capuchins, we notice and are greatly moved by the fact that something is wrong in a world where some get grapes while others get cucumbers. "Let them eat cake" adds grist to the revolutionary mill precisely because of its grotesque unfairness – or its *apparently* grotesque unfairness, if we are stuck with the views of Boehm and Joyce as the best empirical account of morality. And perhaps the appearance of unfairness is all we really need to ground our rallying cries for greater social justice. Still, the better we come to understand ourselves and our moral values in the context of wide reflective equilibria grounded in modern science, the more worrisome it may become to live in a world grounded in moral appearances rather than in moral reality.

We are only stuck with this result if we take something like Boehm's view of conscience to be the ultimate source of moral authority. From a more broadly evolutionary perspective, it is not clear that we should.

First, there is the point about capuchins, which is not so easily dismissed. While capuchins care a lot about fairness, they do not possess any psychological capacity like a conscience. They would thus seem to be responding to unfairness, not the semblance of unfairness created by their consciences. This is precisely what leads many researchers to suppose that capuchins are not really capable of moral behaviour. If moral behaviour requires moral responsibility, we would agree; but if moral behaviour is behaviour responding in a morally positive way to a moral value in the environment, like fairness, the point may go the other way.

So there is some reason to suppose that a deep and pervasive concern with fairness would have already existed in the environment in which Boehm believes our consciences evolved. There would then also be more to having a conscience than just guilt stemming from shame from unsuccessfully trying to run roughshod over the autonomy of others. On our view, conscience is better understood as part of a more complex and human psychological response to negative moral values like unfairness. It is also the mechanism whereby more complex moral rules get internalized as human children morally mature into societies governed by such rules. These internalized rules in turn become crucial inputs to the same ongoing processes of wide reflective equilibrium that produced the original rules. On our view, conscience is an important part of the human relationship to morality but not the sole source of morality, human or otherwise.

A second problem with Boehm's approach is that while he says sympathy is important to the origins of morality, it plays a minimal role in most of his discussion. If Hrdy (2009) is right about the cooperative breeding hypothesis, sympathy is available as an early source of morality and an early source of ongoing cooperation and trust. Boehm tends to see partiality to close kin as nepotism, a kind of favouritism that would be discouraged in more egalitarian hunter-gatherer groups that are trying to divide up valuable food resources more or less equally. But partiality is much more morally complicated than this, and in its earliest forms, alloparenting would have given individuals a direct interest in the interests of others not directly related to them. Alloparenting brings with it the capacity to form commitments to long-term projects, most centrally the ongoing rearing of vulnerable children and the social cohesion required to do this successfully. So, along with a basic concern for fairness, we must also suppose that already in the early human environment there was a direct concern for the interests of others. This suggests that there is much more to the formation of conscience than simply shame and guilt over playing unfairly. Shame is a stick that raises painful bruises, but there is a lot about what causes the human moral skin to bruise that Boehm is ignoring in his account of conscience.

A third and related problem with Boehm's approach comes from Gilbert (2003). Boehm's (2012, 19–20) account of conscience moves very quickly and directly from the idea of guilt to what he takes to be the more inclusive or at least more fundamental phenomenon of shame. In a more psychologically careful effort to disentangle shame and guilt in an evolutionary context, Gilbert distinguishes several forms of shame, and their effect on our sense of self, from guilt, which he says includes a concern for the interests of others. He sees guilt as more closely connected to a care-based ethics of the kind discussed by Gilligan (1982). This would tie guilt more closely to the kind of cooperation discussed by Hrdy mainly regarding co-parenting children than to the kind of cooperation mainly regarding the sharing of meat that is the focus of much of Boehm's discussion of conscience.

These points made, we return to the worry about tyrants. The view defended by Boehm and Joyce allows for the canonization of both Nelson Mandela and Kim Jong-il. While moral truth may not be sufficient, all by itself, to defeat social tyranny, it does seem to be a crucial part of successful efforts to change society for the better, a point we will take up in our chapter on slavery. One might suppose that our moral capacities, all by themselves, could be enough to do this kind of job, without our having to add on the extra topping of moral truth to what we might otherwise be strongly inclined to think to be morally true when it comes to tyrannical social systems. But behind Thrasymachus' worry, there is a further problem with basing morality solely on conscience: once we realize that conscience is simply formed from shame or guilt, we don't really have much of an answer to Glaucon's question of why we ought to be moral. Illusory moral authority, for critically reasonable organisms, is illusory moral authority.

The moral authority of rational self-interest

A less minimalistic view of morality and the authority of moral reasons is to be found in Gauthier's extended contractarian attempt to reduce morality to rational self-interest. This view gets its first full statement in Gauthier (1986), and it receives important revisions in Gauthier (2013). We might see it as evolution's answer to the instability of conscience as the sole source of morality. On Gauthier's view, the authority of moral reasons is not illusory, but firmly grounded in our own rational self-interest, a dominant aspect of human nature. This answer appeals to what we might think of as the lowest common human denominator, because whatever else we might assume about reasonable human agents, we must suppose that they are rationally self-interested. If we could, therefore, reduce morality to rational self-interest, this would seem to put it on as secure a basis as possible in terms of the ultimate authority of moral reasons in our lives.

In an evolutionary context, there are problems with this approach to morality. On the more limited evolutionary perspective of Joyce and Boehm, rational self-interest seems to arrive on the evolutionary scene to solve a moral problem that doesn't exist. If Boehm is right, most of us already had consciences to keep us morally in line. The problems raised by Thrasymachus and Glaucon only seem to arise forcefully once we find ourselves in larger and more stratified social groups, like Greek city states. Self-interest, including rational self-interest, presumably evolved in humans long before that. Why then take rational self-interest to be the source of morality, when we already had consciences that in small groups would have kept most of us from cheating most of the time?

In answer to this question, we might suppose that conscience and rational self-interest co-evolved, with rational self-interest serving as an additional source of self-control in species where individuals were developing critical capacities enabling them to question the wisdom of what others or their own consciences were demanding they do. Rational self-interest and conscience might then be expected to overlap to a considerable degree, with rational self-interest helping out in cases where conscience or the emotional threats of social shame and isolation fail to be persuasive enough guides in telling us what we ought to do. So we might suppose that Boehm's and Gauthier's theories about the origins of morality are not only consistent but mutually reinforcing.

This is clearest if we compare Boehm's approach to the argument in Gauthier (1986). In this book, Gauthier distinguishes between what he calls a capacity for an *affective morality*, what we are here calling a moral capacity focused on the interests of others as well as our own, and an *affective capacity for morality* (Gauthier 1986, 325–329). The distinction is important to him, because if there were a human moral capacity that operated independently of the capacity for rational self-interest, contractarian morality would not be the ultimate answer to the question of why one ought to be moral. On the other hand, there is the problem for contractarianism of what Gauthier, following Hume, calls the sensible knave: the rationally self-interested individual who correctly perceives that whenever cheating is possible, he (or she) should cheat (Gauthier 1986, 182 and 316–317). If we are all knaves at heart and we know this about ourselves and each other, a rational contract is not possible: we have reason to enter into it, but not to keep it. Luckily for us, says Gauthier, we appear to have the emotional capacity to form a conscience, which capacity we may suppose would keep us emotionally committed to the moral contract if it is rational for us to enter into it. This is what Gauthier calls an affective capacity for a rational morality.

Boehm's view of the human conscience seems to be a good fit with this approach to moral truth and moral authority. Given its origins in the sort of shaming and guilt that is likely to arise when we cheat others in cooperative interchanges, it would seem likely to be plastic enough to form itself neatly around the deeper contractarian moral truth about how we ought to act towards one another based on rational self-interest.

Gauthier's contractarian focus is on solving prisoner's dilemmas, one-off cooperative exchanges where cheating is possible. Although they are important, prisoner's dilemmas hardly exhaust the cooperative landscape. In the early human environment, a lot of cooperation would have already been in place that would not have had this structure. In addition to the kind of sharing behaviour that is at the centre of Boehm's approach to conscience, there is again Hrdy's (2009) important work on alloparenting, where caring about the interests of others would have been much more significant than either having a conscience or being rationally self-interested. Alloparenting starts in species a long way away from either of these more complicated human capacities or even their possibility. We were most likely cooperative breeders before we were hunters and gatherers or crafty practical reasoners confronted by prisoner's dilemmas. Hence, we were never in an evolutionary context where we approached prisoner's dilemma problems as purely rational agents, and it is thus unlikely that Gauthier's purely rational solution to prisoner's dilemma problems is at the foundation of all human morality. We most likely evolved to have a moral capacity, and it seems unlikely that this capacity is entirely consistent with the dictates of rational self-interest, unless perhaps the selfish gene hypothesis of Dawkins (2006) is correct, a point we take up later in this chapter.

In the meantime, there are other problems with contractarianism. Even if there is more to morality than contractarianism takes into its scope, Gauthier claims that contractarianism gets right the core bit of morality that involves cooperative interchange or at least those cooperative interchanges that can be modelled by games like the prisoner's dilemma. This leaves open just how core this form of cooperation is to morality, and how this core bit of morality is related to other, perhaps equally important, bits of morality. It may be that morality is not a well-integrated phenomenon, and so perhaps some of its dictates may be contradictory with one another with no morally appropriate way of resolving such contradictions. But then, why not suppose that morality is also connected to reason, or at least to that part of reason which is connected to the search for coherent forms of wide reflective equilibrium? If this were so, the rational social contract of contractarian moral philosophy would not be the ultimate source of moral reasons and moral authority, although rational self-interest might still be supposed to be an important part of morality.

The purchase that contractarianism has on contemporary thought comes at least in part from the fact that those of us living in economically developed Western countries are ideologically immersed in increasingly strong forms of liberal individualism. But liberal individualism does not presuppose or require contractarianism to be true, and contractarianism is hardly a dominant approach in contemporary Western ethics. While Western arguments regarding moral issues take into account self-interest, this is not typically their most interesting or significant feature. Consider briefly Wilkinson and Pickett's (2010) argument, based on extensive epidemiological research, that as economic inequality increases in developed societies, health and well-being are diminished for each sector of society in fairly even gradients from the top to the bottom economic percentiles. Even those individuals in the top percentiles are better off in societies were the economic gap is smaller rather than larger. The top 10% of Swedes, for example, have a lower infant mortality rate than the top 10% of those in Britain. What seems most striking about these data, however, is not the idea that in whatever class we might happen to be, the risks to our health and well-being would be less than if we were in a more equal society, but rather that it is manifestly unfair that morbidity and mortality should be distributed in such a direct parallel with economic advantages and disadvantages. People might deserve more or less wealth or income, but they hardly deserve, on this very basis, more or less morbidity or mortality. This is just not fair. But the contractarian might still point out that the moral problem he or she is trying to solve is the deeper one raised by Thrasymachus and Glaucon. As we noted in discussing Boehm, Thrasymachus wonders if justice is nothing more than an homage the powerless must pay to the powerful, and Glaucon wonders more deeply about the irresistible appeal of the mythical Ring of Gyges, a magic charm that would enable its wearer to become invisible and thus have power over all others. As we have already noted, tyrants can arise. This was possible in Greek city states, and it is possible in modern states as well. It is not clear, though, that contractarianism has a response to this problem.

Gauthier (2013) suggests that though the path of the tyrant might look individually rational, it isn't really, once we take into account that any rational act must appear rational to all would-be cooperators. If I'm rational, I must take into account that all those I might cooperate with are rational as well and that cooperation is always likely to yield more dividends than not cooperating. A similar line of argument might be used to account for the very real unfairness of the health outcomes we considered earlier: while large gaps in wealth and income may seem justified on purely economic terms, we can't expect rational social contractors to accept such gaps if they are tied in systemic ways to significant differences in mortality and morbidity. Rational cooperators might accept uneven economic pay-offs if they mean more money for everyone, whatever economic percentile he or she might wind up in, but it seems irrational to accept systemically higher rates of morbidity and mortality for oneself and one's children based on wherever one winds up economically in a modern liberal democracy. The inequalities in health and well-being that Wilkinson and Pickett draw our attention to are indeed unfair, and contractarianism tells us exactly why.

Gauthier (2013) can thus let go of the 1986 argument that rationality requires emotions to buttress its normative authority in moral contexts. Moral norms are supported solely by the normative authority of rational self-interest, properly understood. In entering into a social contract, rational cooperators must seek to bring about a Pareto-optimal result whose pay-offs are rationally acceptable to all those who are parties to the contract. The basic idea is that if we all know that we (both ourselves and others) are rational, we must all see that what we are told to do by rational self-interest has to be rational from the point of view of all parties to the contract, insofar as all of us, as parties to the contract, are expected to be equally rational. This idea leads to what Gauthier (2013) calls the contractarian test for whether a given set of social arrangements is justifiable from a moral (i.e., rational) point of view. For any particular social rule or institutional arrangement, the test requires us to ask whether it is the sort of rule or arrangement each of us would agree to, as rational individuals seeking rules that will jointly govern our actions towards one another. Rules that tell would-be knaves to cheat when they rightly think they can get away with it do not pass this test, and so knavery is ruled out.

The problem for this version of contractarianism is that social contracts can fail to be fully inclusive. As argued in Pateman (1988), Mills (1997), and Pateman and Mills (2007), social contracts may be deeply biased by gender, race, and other attributes of people that can be used to divide them into separate social classes. According to these arguments, this has been an ongoing practical problem not just within Western democracies but, more deeply, within the Western theories of social justice that are held to legitimate these democracies morally. Because of the ways in which our thinking about each other may be biased by such things as sexism or racism, rights and liberties we understand to be universal turn out not to be so, not for all classes of people. In an extended example in Pateman and Mills (2007, 35–78), Pateman discusses at length what she calls the original settler contracts of the fledgling democracies of the New World. Eschewing earlier justifications of the subjection of native populations based on the concept of conquest, political thinkers like Grotius and Locke argued that because New World land was not under private cultivation and because the people using this land had not organized themselves into political states to protect individual liberty and private property, the land was *terra nullius*, meaning the New World was still in a state or nature and hence open to the development of an original rational contract that would establish both private property and legitimate political sovereignty. Pateman summarizes the argument as follows:

> In a *terra nullius* the original contract takes the form of a *settler contract*. The settlers alone (can be said to) conclude the original pact. It is a racial as well as a social contract. The native peoples are not part of the settler contract – but they are henceforth subject to it, and their lives, lands and nations are reordered by it.
>
> (56)

According to Gauthier's contractarian test, such exclusionary institutional arrangements would appear to be ruled out by the assumption of shared rationality: if it is not rational for you to accept your position in society, it is not rational for me to expect that I can impose it on you. But it is not clear that such institutional relationships really are contrary to rational self-interest. For those on the positive side of such contracts, these institutional arrangements can pay off (and have indeed paid off) quite handsomely. As Pateman and Mills argue, it is all too easy to hide the true nature of such arrangements for centuries. Given the nature of social power and the social situations of those who wield this kind of power, it may be rational not to question it, to the extent that it can even be credibly identified as such, but instead, to get the best deal you can within the sexist and racist constraints of the society you find yourself within. Given our gender or our race, it might be best for us, whoever we are, to agree universally to sexual and racial social contracts, that is, social contracts that systemically and to some degree invisibly discriminate against certain classes of individuals.

From Singer's pure-reason perspective, we might argue that the moral problem with exclusionary social contracts is that they are logically inconsistent. But this misses the point that the kind of inconsistency at stake here has a distinctively moral aspect to it: the problem with exclusionary social contracts lies more plausibly in the fact that they are simply not fair. Before there were laws of logic or at least before we humans were around to discover them, there were fair and unfair kinds of social interactions among species of animals with varying levels of intelligence and mutual dependence. Recognizing that someone is being treated unfairly does not seem to require being able to reason in a logical manner. Moreover, logical inconsistency can itself be a moving moral target; we are us, in any exclusionary contract, and they are them. Why isn't this the most morally important thing about us and them, namely that we are us and they are not?

We take up the us/them problem from an evolutionary perspective in Chapter 7. We conclude our discussion here with the point that to the degree that we enlarge the concept of rational self-interest itself to include the further idea that exclusionary social contracts fail to treat others as we would ourselves like to be treated, we push contractarian thinking in the direction of contractualist thinking. For the contractualist, social rules and institutions must be justifiable to each individual insofar as he or she is rational, subject to the further constraint that we must limit the regard we would hold ourselves in by an equal regard for all others with whom we would or could enter into a social contract.

Before moving on to consider contractualism, we should note that EMR need not suppose that rational self-interest and morality are unrelated, but only that morality does not reduce to rational self-interest, which is the key claim of contractarianism. Given the kind of things that moral values are, according to EMR, we might expect that paying attention to them would often be in an individual's rational self-interest. For the evolutionary success of the human species, this is a fortunate thing, because if rational self-interest and morality pulled too often and too hard against one another, we likely would not have evolved as the kind of species that we are.

Contractualism as a source of objective moral truth

Contractarian theories like those of Gauthier (1986, 2013) look back to the kind of social contract theory first developed by Thomas Hobbes. If each individual human agent is rationally self-interested, will society be a good thing for each individual, and if so, what sort of society will be best for each? For Hobbes, the answer to these questions was a society that guaranteed peaceful coexistence through the mechanism of a powerful Leviathan who prevented everyone else from cheating the system. Most modern contractarians want to avoid this result, but like Hobbes, they still want to reduce morality to some version of rational self-interest.

Contractualist theories, on the other hand, look back to the moral and political philosophy of Kant, and ask what sort of social contract would be selected by rationally self-interested human agents who considered themselves to be moral equals. To put the point in explicitly Kantian terms, if you and I are equal rational agents, each of us needs to recognize that the rationally chosen ends of the other are just as important to him or her as our own rationally chosen ends are to us.

Salient among contemporary contractualist approaches to morality are Rawls (1971), Scanlon (1998), and Copp (1995, 2007). Strictly speaking, Rawls (1971) does not belong to this list, since his 1971 theory of justice is a theory of social justice rather than a theory of morality. We include it here because it is a well-known contractualist theory and because it provides an especially clear example of the combination of rational self-interest and equal regard for others that is the hallmark of contractualist theories of morality.

Rawls famously asks what principles of social justice individuals would choose for themselves in a position abstracted from their current set of social relationships. This choice situation he calls the Original Position, a position where individuals know they are in a society with different roles and rewards but where no individuals know what their role and rewards might be, and crucially what their own talents and abilities might be. From behind this Veil of Ignorance, individuals will be forced to choose principles of social justice based only on rational self-interest and equal regard for others. Although each will want to do as well as possible for himself or herself, no one will know what his or her individual talents or role in society might be. Faced with this choice, individuals will choose, argues Rawls, basic social institutions that distribute social and political power in ways that are equally open to all who might compete for them, and then wealth and income will be distributed in such a way that those who are least advantaged will receive as much as they possibly can. Differences in wealth and income are justified only if they benefit to the greatest degree possible those with the least.

Our point here is not to explore the details of Rawls argument but to register that on this argument social justice is defined by, and hence reduced to, the choice individuals can be supposed to make in the Original Position, a choice made solely on the basis of rational self-interest and equal regard for others. The mechanism of equal regard is the Veil of Ignorance, which ensures that no one can prioritize their individual ends over the ends of anyone else in choosing the basic structure of society because no one knows what in fact their ends will be. A key point for Rawls is that the principles of social justice so selected will not be utilitarian in form, because in some situations utilitarian principles can be used to justify sacrificing the ends of the few to realize the ends of the many. Not knowing who we are ensures, thinks Rawls, that no one of us would ever agree to this sort of result, where individual goods are aggregated across persons and traded off, one set of such ends against another. As individuals with equal regard for one another as separate persons, each with his or her own ends, no one of us would want our own good lumped together with the goods of others and then traded off against other similar sets of aggregated goods. Such trade-offs would fail to respect us as the separate but equal persons that we all are.

Scanlon (1998) applies this same sort of contractualist argument, with a similar non-aggregationist result, to morality itself. Scanlon's general claim is that we are not simply rationally self-interested creatures and that questions about what we owe to each other form a central part of human morality. What we owe to each other, Scanlon argues, is determined by what we can reasonably require from one another; the contractualist test of a moral principle or rule, for Scanlon, is whether it can reasonably be refused by anyone who would be affected by its application in human affairs. The notion of "reasonable refusal" is the lynchpin of Scanlon's contractualism, and he develops it carefully and insightfully. In terms of our argument here, two of his claims are key: what it is reasonable to refuse is not limited or otherwise determined by what it is in anyone's rational self-interest to refuse, and what it is reasonable to refuse, all individuals and all things considered, cannot lead to aggregationist moral principles like those of utilitarianism. Against welfarist versions of utilitarianism, Scanlon does not suppose that individual welfare is the ultimate moral good that provides the measure of all other derivative forms of moral goodness. More generally, against all versions of utilitarianism, he does not think aggregate forms of moral goodness are necessarily decisive in telling us what we ought to do.

An illustrative example may be helpful. Scanlon thinks that sometimes individual goods can be aggregated or at least appear to be aggregated. Consider a principle of rescue that tells us that when a larger and smaller number of individual lives are threatened, we can save either those individuals in the larger group or those individuals in the smaller group (Scanlon 1998, 240). Scanlon argues that this principle fails to pass the contractualist test. Consider the case where the larger group contains two individuals and the smaller group contains one individual. Either one of the two individuals in the larger group can reasonably argue against the principle that it gives no weight to his or her life, because the principle treats the two groups as if they were of equal moral importance. If each human life is equally important, the group containing the one person and the other person in the two-person group have their lives given equal weight if the would-be rescuer can choose which of either group to save, but in this case, the second person in the two-person group has had the importance of his or her life entirely left out of moral consideration. Thus, the second person in the two-person group can reasonably refuse to agree to the principle of rescue suggested earlier and it cannot be accepted as telling us what we owe each other when it comes to rescuing groups of people from harm.

To some extent, EMR can remain agnostic on this issue, and more generally on contractualist approaches to morality. Contractualist thinking may be an important aspect of human thinking about human morality. EMR's point is that this is not all there is to morality, and that human moral thinking should also take natural moral values into account as we reason towards consistent and coherent systems of moral judgments at all levels of generality, from particular cases to moral rules to moral principles. On the other hand, humans are sometimes swayed by utilitarian sacrifices of the interests of the few to benefit the interests of the many, and to the extent that we are, this suggests that there is more to the human moral sense than is captured by the contractualist's notion of reasonable refusal and the deeper moral regard for each other as separate individuals that it rests upon.

Copp (1995, 2007) provides another contractualist account of morality, one that is explicitly reductionist and naturalistic. First, whether or not a principle can be reasonably refused is ultimately a matter of objective fact. So, Copp's contractualism gives us an account of moral truth: a moral rule or principle is true if and only if it is justifiable to others in the right sort of contractualist way. Second, Copp argues that moral rules and principles that cannot be justifiably refused are tied directly to natural facts about what best satisfies basic human needs. So, moral truths are ultimately natural truths, and Copp's contractualism thus gives us a reductionist version of moral naturalism. There are no real moral values as such, but there are moral truths, and these moral truths are based on natural facts about basic human needs and how best to satisfy them. Again, EMR need not deny all aspects of this view but only its reductionist account of moral truth. Contractualism may be tracking an important aspect of human morality without its being a complete account of human morality or morality more generally.

Contractarianism and selfish genes

A main attraction to thinking that contractualism is tracking an important aspect of human morality is in the way in which it combines human rationality with human emotional commitments to fairness and impartiality. In the next several chapters, we want to begin to tie these attractive features of contractualism to the natural moral values and adaptive capacities discussed in the first part of this book. But before we do so, we need to deal with some unfinished business from our last chapter that is directly related to what might appear to be an attractive fit between Gauthier's (1986) contractarianism and Dawkins' (2006) selfish gene theory. This tangent to the main line of argument in this chapter will prove lengthy, but it will provide some important empirical detail to EMR's understanding of moral values as natural kinds. This detail matters to our argument that human morality is one particular instance of a more general biological phenomenon, and hence the natural kinds we discuss in the first half of the book are in fact moral kinds.

Contractarianism seems to offer a tempting picture of the small human finger of morality reaching out to touch the much larger biological finger of selfish gene theory, with something of the same aura of inevitability of Michelangelo's painting in the Sistine Chapel. According to Dawkins, the first fully biological individuals to appear in the natural world were genes: individual entities competing with other individual entities over resources and relative reproductive success. As individual genes combined with other genes to form strands of DNA, these strands of DNA combined to form organisms which themselves appeared first and foremost as biological individuals competing with each other, once again over resources and relative reproductive success. A central puzzle for Dawkins is why such organisms would cooperate with one another where such cooperation would require one organism to sacrifice its interests in favour of another. Why do selfish genes simply not build selfish organisms?

The two main answers Dawkins offers to this question are well known in the biological literature: kin selection and reciprocal altruism. If an organism sacrifices its individual interests for the interests of a close relative, more copies of the genes these two individual organisms share may end up in future generations, thus ensuring greater reproductive success for the genes of each individual organism. In reciprocal altruism, one organism benefits another at an immediate cost to itself to increase the likelihood of future benefits that will repay such costs. Given the relatively small immediate costs, organisms that cooperate in reciprocal ways may reap greater benefits than if they did not cooperate in these ways. Like kin selection, reciprocal altruism may thus be an avenue to greater inclusive fitness for the organisms that behave in this general kind of way.

The main problems with this picture are that genes are not as atomistic as the selfish gene picture requires, nor are humans as individualistic as reciprocal altruism assumes. Although kin selection and reciprocal altruism can by themselves do a lot of moral work, moral sense theories, and EMR as well, suggest that there is much more going on in the evolution of morality than can be accounted for by just these two factors. But at a more fundamental level, there are three significant problems with the selfish gene theory: first, genes do not work alone but work together in complex and perhaps irreducible ways to produce traits; second, traits are not individually selected in many if not most cases; and third, group selection in certain sorts of creatures is at least probable if not fully proven. We will consider each of these issues in turn.

As Evelyn Fox Keller (2000) argues, the general neo-Darwinian ideas that genes determine traits, variations in genes determine variations in traits, and selection processes acting on traits select the genes that produce these traits are based on a naïve understanding of how genes are expressed. There can be many–one, one–many, and many–many relations between functional genes (those that express proteins) and traits. Much of this occurs through manipulations of mRNA before its information is carried to ribosomes. Exactly this sort of complexity has become recently and strikingly evident in cephalopods (Liscovitch-Brauer et al. 2017), where changes in the mRNA are the main drivers of evolution and the DNA must remain fairly constant or the RNA variants would not be able to function.

In general, genes do not work alone to produce proteins and eventually traits, but they interact with each other and their environments in complex ways. If we start with the standard picture of a double helix of DNA strands, gene expression begins when a particular strand of DNA is copied into a single strand of RNA that provides the information for protein synthesis. This strand of RNA may fold onto itself (creating new biochemical structures) and it may interact with other RNA strands and proteins in the cell. Depending on how the RNA strand has folded itself or interacted with other RNA strands or proteins, the RNA strand will produce more or less of certain kinds of protein molecules, which will themselves have different effects on phenotypic trait development. Folding may be caused by other strands of RNA, features of the original RNA strand itself, or other biochemical factors such as proteins inside or outside the cell in which these processes are occurring. It is unclear whether such processes can be reduced to the additive effects of the components of a DNA, RNA, or protein sequence. There may well be additional information that emerges from the folding. Whether there is or not we do not currently know, but it is possible and it is even likely if the self-interactions of RNA in a typical cellular environment are organized along the lines of complex organization described by Collier and Hooker (1999).

Much successful neo-Darwinian evolutionary research has been on the selection of specific traits that can be related to specific genes. However, in addition to the problems just mentioned, traits themselves are not selected in isolation but in the survival or not of the carriers of the genes, which involves many co-evolved traits. Given the possible irreducible complexity of the organization of complexes of traits, it may well be impossible to select some traits individually, even under the best of experimental conditions. Furthermore, selection of complexes of traits may occur through a process of "tuning," the rhythmic entrainment of complex aspects of the environment with the complexity of organisms and groups of interacting organisms (Collier and Burch 2000; Collier 2007). Tuning occurs when the component parts of a complex system fall into a stable way of interacting with one another that may or may not be an optimal mode of functioning for each of the individual component parts. In other words, in some tunings, some component parts might function more optimally were the system to be tuned to some other stable state. But once a system is already tuned in the way that it is, it may take more energy to move the system from its current tuned state to another more optimally tuned state. In such cases, the reduction of selection to genic selection would be impossible because what is best for a system based on its parts will not be a computable function (Collier 2008, 2010). In general, evolution by natural selection does not imply reducibility, however tempting such origin stories might appear to us. In the next section, we will go on to argue that at the level of trait selection, moral capacities are likely to be grounded in non-reducible instincts such that there is again no computable algorithm for matching environmental inputs to behavioural outputs.

Finally, it can be argued that group selection was ruled out as a likely biological explanation much too quickly, and, moreover, selection for group properties may well yield moral adaptations (Sober and Wilson 1998; Wilson and Wilson 2008). The worry about group selection is that if it is required for cooperative behaviour that goes beyond self-interest, a not unreasonable assumption, then a group of cooperators would likely be invaded by cheaters who take but do not give. In general, the idea that groups are selected for group traits rather than merely individuals for individual traits violates common reductionist accounts of the selection of traits. There are at least two problems here. First, Nowak, Bonhoeffer, and May (1994) showed that in a simple model of cooperators and cheaters spread in two dimensions, where cheating by cooperators gives a higher fitness than cooperating, mixtures of cooperators and cheaters can be maintained indefinitely. So, it is not clear the invasion hypothesis is true. This model does not lead to the dominance of cooperators, however, which we require for EMR to be a reasonable explanation of the generality of morality. If punishment is introduced, if its cost is small enough, punishers who punish (or just isolate) cheaters will thrive. There is a new version of the invasion problem at this level: punishment is risky and costs the punisher, so it might be better off for each individual to avoid punishing and hope that another member of the group does it. If, however, clusters of cooperators are relatively stable, and punishers are also not wiped out by free riders by similar dynamics, it is possible that groups that have a biological propensity for cooperation will survive where clusters of cheaters would perish, that is, there will be group selection for cooperative behaviour.

This process can be strengthened if there is also a propensity for punishing cheaters caused by the adaptations of the group. Something like this can get started by kin selection for cooperating with kin and punishing those who hurt kin. In fact, we know that kinship is important in explaining caring for others. Selection, contrary to a common assumption among some biologists, is actually seldom optimal, and kin selection might lead to the selection for care in a more general way, perhaps because checking for kinship is too complicated or just because less discriminating care yields enough of an advantage to kin to be selected for even if the care involved is not fully kin oriented. Group selection processes could then act on groups whose members care or punish in the ways we are imagining. Ironically, perhaps, if cooperation with other cooperators and punishment of cheaters become dominant in a group through group selection, then it may well become advantageous for the individual to conform to this organization, on pain of punishment and loss of access to resources that are best produced by cooperation. Individual selection and group selection can mutually reinforce each other under the right conditions.

What are the right conditions? Typically the groups must be small and independent for group selection to be effective. A paradigmatic example is myxomatosis in rabbits. Groups of the virus that are highly virulent tend to die out because they kill their hosts too quickly for them to spread to other hosts, so less virulent forms tend to be more successful and eventually take over. Humans and a number of other animals discussed in earlier chapters form relatively small bands that are relatively isolated. These would be ideal subjects for group selection.

Moral instincts redux

Tied up with the selfish gene theory is the idea that something as apparently complex as morality might be reducible to moral behaviour patterns that might themselves be further reducible to psychological mechanisms that are optimized to maximize genetic self-interest, whatever this term is supposed to mean. In the comparative study of animal behaviour, appeals to instinct may seem dubious because of early proposals regarding the nature of instincts that proved to be highly problematic (e.g., Lehrman (1953)). But along with others who wish to rehabilitate the term, we cannot think of a better word to refer to open-ended behaviours related to sex, finding nutrition, language, and of course morality. Whereas behaviourism deals only with behaviours and their conditioning, ethologists for some time have used instincts to explain complex behaviours that are not easily reducible to stimuli, reflexes, and conditioning.

To develop EMR further, we would adopt Jean Piaget's (1971) account of instinct because it avoids some of the earlier problems in defining instincts. Although Piaget originally believed that reflexes and conditioning were sufficient to explain intelligence in terms of assimilation and accommodation, he eventually gave this up, realizing that open-ended schemata rather than the closed schemata derived from conditioning behaviours alone were needed to explain intelligence. Along with comparative psychologists like de Waal, we think EMR should suppose that similarly open-ended schemata or instincts are required to explain morality.

For Piaget, psychological operations on received stimuli can yield open-ended behaviour patterns or, to use his (1971) term, schemata. Schemata are basically patterns involving relations between stimuli and psychological operations on them that underlie organized behaviour. Instinctive schemata differ from their behaviourist counterparts in being open-ended. Behavioural schemata are closed, with well-defined ranges of inputs and outputs. Behaviour related to nutrition, for example, can include a wide range of possibilities that are not predetermined by a specific set of inputs, although nutrition habits, like other instinctive behaviours, can be conditioned and can vary culturally.

Against Piaget, a reductionist might take the view that instinctive behaviours can be theoretically reduced to the brain activities connecting inputs and outputs. There is reason to believe, however, that although there are appropriate brain causes, they are not always the same for the same instincts. One of the consequences of divergent phylogeny coupled with convergent schemata types is that the latter may have widely differing neurological bases. This is especially evident between birds and mammals across sound production syntax, tool use, and social behaviours, because the lineages became separate before the emergence of dinosaurs. This also appears to be true of moral capacities. It seems that similar schemata can stem from significantly different neural structures. We conclude that not only have attempts to find explanatory closed behavioural schemata for instinctive behaviours failed, but also that there do not appear to be common neural correlates either. This is a problem for the Wild Justice view of Bekoff and Pierce, but not for EMR. EMR's moral kinds depend on sameness of structural features in particular environments, not on the sameness of neurological causes in heads coupled with an apparent sameness of behavioural consequences in the world.

Biologically, both stimuli and responses are patterned, with a further pattern relating the two. The relational pattern is functional in nature. Instinctive schemata arise because of some functional match between available means and ends. The matching relation is an isomorphism at some level between the environment and the goals of the organism.

Towards suggesting a more dynamical account of how such matching can arise, we make several observations. The environment and organism and their structural relations are all complexly organized. Presumably this is also true of combinations of the relevant parts involved in an instinct, whether they are components of the instinct, the composition of its physiological building blocks, or its input–output relations. The components are also, of course, dynamical entities. Complex organization implies (likely) irreducibility of wholes to parts because it involves non-local interactions through feedback and feedforward processes, both of which are recognized sources of irreducibility in even some very simple systems. This non-locality is similar in nature to the example given for genes in the previous section, although it involves the neural processes as well as possibly the interactions with the world, which are also often circular in nature. The isomorphism that Piaget invokes is grounded in a dynamical congruency: feedback can be reinforced, and it can be induced by feedforward. Complexly organized systems with similar dynamics can become tuned to one another to produce a congruency, allowing systems to interact without being reducible. This can produce irreducible schemata that are open-ended.

A dynamical approach to moral instincts may allow for a better account of how such instincts evolve. Because of the possibility of tuning in complex systems, instinctive behaviour can evolve through selection processes acting on tuned behaviour. Tuning can occur through behavioural modification and then selection can act on the traits that tend to produce this modification. Just as instincts can enter into conditioning and habits, they can also enter into selective processes.

Such processes would possess a general characteristic of interactions within complex systems, like the imposition of some force or a constraint (such as an observation or other stimulus). When a force acts on a complexly organized system, it tends to modify itself so as to reduce the effect of the force in the direction of the applied force. This leads to orthogonal forces, producing a cascade that modifies the system as a whole so as to accommodate the force (Collier 2001). There is no particular reason to think that this is restricted to complex physical and chemical systems, and Collier (2001) applies the idea of orthogonal forces to give a dynamical account of Piaget's processes of assimilation and accommodation. Once the modifications are in place (accommodation), selection can adapt to the external constraint or force.

This sort of process for adaptation is possible only in complexly organized systems. The organism is active in the process, unlike in random variation and selection in which the organism is a passive mechanistic component. Furthermore, the process is irreducible, or emergent, as discussed in Collier (2008, 2014), when the rules governing the system interact with the boundary conditions. The modification of the instinct depending on environmental conditions ensures this, unlike in the case of behaviourism.

Finding behaviourist explanations and cases of behaviour that might otherwise be supposed to be instinctual is not sufficient to reject irreducibility because instincts can be conditioned and become habits. The open-ended character of social interactions, language, altruistic behaviour, tool use, and a host of other phenomena across lineages suggest convergence on common schemata found in functional isomorphisms due to common problems to be solved. This requires, if we are right, operational interaction between behaviours, internal states, and the environment in ways that cannot be separated, producing partially closed loops of feedback and feedforward. In particular, we should be looking for cases in which motivational processes and perception of stimuli cannot be easily separated. Best would be an explanation in terms of information loss and corresponding self-organization within the organism.

Assuming that instincts play a role in the creation of moral behaviours, along with other general and cross-species complex behaviours, what will an explanation of behaviour in terms of instincts look like? First, complexly organized instincts allow matching between open-ended environmental properties and behaviour. Piaget believed that our intellectual development follows the same pattern as biological evolution, and the same may be true for moral development. Instincts can ground further cognitive adaptation, whether by selection or by behaviourist reinforcement. This can create the conditions required for matching of more complex innate tendencies with the environment, producing levels of instinctive behaviour, interconnected by learning processes. Instincts and the way they are expressed are not entirely path and organism independent but show some degree of both sorts of dependence, something found in complexly organized systems more generally. The important thing here is that the adaptation by instinct depends as much on the type of creature as on the particular evolutionary history and is thus more like an adaptation to moral kinds than a species-specific trait.

Moral kinds

Collier (1996) argues that natural kinds are needed for scientific explanations. We need to describe relations between general properties in science, and these relations must be necessary in some sense to give them scientific force. Collier (1996) argues that this is possible if kinds are grounded in particular causal relations where the causal structure has elements in common across different instances. The necessity is not of the sort existing across all possible worlds, because scientific laws are contingent. What is necessary is that in any relevant world in which instances of the general structures exist, the relations are the same. These structures are natural kinds.

For EMR, moral capacities are adaptations to moral natural kinds, in Collier's sense of the term. The social organization and potentials of a suitable group provide the environment in which we can get a matching between social environment and moral capacities through the process of natural selection which requires a matching of conditions favourable for moral-like behaviour and biological processes grounded in inherited characteristics. This matching requires an appropriate matching of kinds. So moral kinds, we think, result from general causal relations between particular kinds of creatures and their social environments. The kinds therefore can be general across unrelated species and need not be recognized in any conscious way to be causally effective.

Philosophers typically want to reserve the term "morality" only for cases where there is reflective awareness of choices made in behaving in one way or another. This sort of autonomous and reflective awareness seems to be required for moral responsibility, which is a component of human morality as it is generally understood; however, as we have argued in earlier chapters, adaptation to moral kinds seems possible with only minimal or even no conscious awareness. Whether we call this proto-morality or some similarly qualified term, there is a general sense in which there is a sameness of kind that underlies all such adaptations. By studying these adaptations, we can get a better understanding of the nature of morality in general and of its biological basis as the natural kind of thing that it is. Human morality then emerges as one possible realization of this same general kind.

If morality starts with natural moral kinds, it is unlikely to reduce to any sort of lower level biological phenomena in any theoretically interesting sort of way. That moral instincts arise as adaptations to natural moral kinds makes the reducibility of a biologically based morality to something non-moral even less likely than we might otherwise suppose. Given the variability of the social environments in which they arise, moral natural kinds are unlikely to be realized in exactly the same sorts of ways across such environments, and given the variable importance of these kinds of things to the various species in question, the moral instincts that are selected for are unlikely to match up with the moral kinds in exactly the same kinds of ways. Helping behaviour in dolphins and chimpanzees involves different physical and social environments and most likely different instincts with different neurophysiological bases. What counts as helping another individual is likely to vary depending on whether a species' environment is aquatic or arboreal, as is the general Umwelt of the species in question. Adaptations just have to be good enough, not optimal, and even then what is good enough is a moving target due to other adaptations, environmental change, and the difficulty of producing particular adaptations that are biologically/genetically accessible given the species' developmental history. If there is a biological basis to morality, it is unlikely to reduce to behaviour patterns that maximize the reproductive success of particular genes.

EMR thus claims that there is more to morality than human morality, and moreover, that the natural moral values that are part of morality as it appears through evolutionary processes in the natural biological world are likely to play a key role in the development of human systems of morality. It is to the defence of this second claim that we now turn.

Bibliography

Baier, Kurt. 1958. *The Moral Point of View*. Ithaca: Cornell University Press.

Boehm, Christopher. 1999. *Hierarchy in the Forest: The Evolution of Egalitarian Behaviour*. Cambridge, MA: Harvard University Press.

Boehm, Christopher. 2012. *Moral Origins: The Evolution of Virtue, Altruism, and Shame*. New York: Basic Books.

Brosnan, Kevin. 2011. "Do the Evolutionary Origins of Our Moral Beliefs Undermine Moral Knowledge?" *Biology and Philosophy* 26 (1):51–64.

Collier, John. 1996. "On the Necessity of Natural Kinds." In *Natural Kinds, Laws of Nature and Scientific Reasoning*, edited by Peter Riggs, 1–10. Dordrecht: Kluwer.

Collier, John. 2001. "Dealing with the Unexpected." Partial Proceedings of CASYS 2000: Fourth International Conference on Computing Anticipatory Systems, *International Journal of Computing Anticipatory Systems* 12: 212–221.

Collier, John. 2007. "Rhythmic Entrainment, Symmetry and Power." In *Explorations in Complexity Thinking: Pre-Proceedings of the 3rd International Workshop on Complexity and Philosophy*, edited by Kurt A. Richardson and Paul Cilliers. Mansfield, MA: ICSE Publishing.

Collier, John. 2008. "A Dynamical Account of Emergence." *Cybernetics and Human Knowing* 15 (3–4):75–86.

Collier, John. 2010. "A Dynamical Account of Individuation and Diversity." In *Complexity, Difference and Diversity*, edited by Paul Cilliers, 79–93. Berlin: Springer.

Collier, John. 2014. "Emergence in Dynamical Systems." *Analiza i Egzytencja (Analysis and Existence)* 23:17–40.

Collier, John, and Mark Burch. 2000. “Symmetry, Levels and Entrainment.” *Proceedings of the International Society for Systems Sciences*, Toronto.
Collier, John, and C.A. Hooker. 1999. “Complexly Organised Dynamical Systems.” *Open Systems and Information Dynamics* 6:279–331.
Copp, David. 1995. *Morality, Normativity, and Society*. Oxford: Oxford University Press.
Copp, David. 2007. *Morality in a Natural World: Selected Essays in Metaethics*. Edited by Jonathan Lowe and Walter Sinnott-Armstrong, *Cambridge Studies in Philosophy*. Cambridge: Cambridge University Press.
Dawkins, Richard. 2006. *The Selfish Gene (30 Anniversary Edition)*. New York: Oxford University Press.
Enoch, David. 2011. *Taking Morality Seriously: A Defense of Robust Realism*. Oxford: Oxford University Press.
Gauthier, David. 1986. *Morals by Agreement*. Oxford: Oxford University Press.
Gauthier, David. 2013. “Twenty-Five On.” *Ethics* 123 (4):601–624.
Gilbert, Paul. 2003. “Evolution, Social Roles, and the Differences in Shame and Guilt.” *Social Research* 70 (4):1205–1230.
Gilligan, Carol. 1982. *In a Different Voice: Psychological Theory and Women's Development*. Cambridge: Harvard University Press.
Harman, Gilbert. 1977. *The Nature of Morality: An Introduction to Ethics*. New York: Oxford University Press.
Hrdy, Sarah Blaffer. 2009. *Mothers and Others: The Evolutionary Origins of Mutual Understanding*. Cambridge, MA: Harvard University Press.
Joyce, Richard. 2006. *The Evolution of Morality*. Edited by Kim Sterelny and Robert A. Wilson, *Life and Mind: Philosophical Issues in Biology and Psychology*. Cambridge, MA: The MIT Press.
Kahane, Guy. 2011. “Evolutionary Debunking Arguments.” *NOUS* 45 (1):103–125.
Keller, Evelyn Fox. 2000. *The Century of the Gene*. Cambridge: Harvard University Press.
Lehrman, Daniel S. 1953. “A Critique of Konrad Lorenz's Theory of Instinctive Behavior.” *Quarterly Review of Biology* 28:337–363.
Liscovitch-Brauer, Noa, Shahar Alon, Hagit T. Porath, Boaz Elstein, Ron Unger, Tamar Ziv, Arie Admon, Erez Y. Levanon, and Joshua J.C. Rosenthal. 2017. “Trade-Off between Transriptome Plasticity and Genome Evolution in Cephalopods.” *Cell* 169 (2):191–202.
Mackie, John Leslie. 1977. *Ethics: Inventing Right and Wrong*. Harmondsworth: Penguin.
Mills, Charles. 1997. *The Racial Contract*. Ithaca: Cornell University Press.
Nagel, Thomas. 1974. “What Is It Like to Be a Bat?” *The Philosophical Review* 83 (4):435–450.
Nagel, Thomas. 1986. *The View from Nowhere*. Oxford: Oxford University Press.
Nowak, Martin A., Sebastion Bonhoeffer, and Robert May. 1994. “Spatial Games and the Maintenance of Cooperation.” *Proceedings of the Natural Academy of Sciences* 91:4877–4881.
Pateman, Carole. 1988. *The Sexual Contract*. Stanford: Stanford University Press.
Pateman, Carole, and Charles Mills. 2007. *Contract and Domination*. Malden: Polity Press.
Piaget, Jean. 1971. *Biology and Knowledge: An Essay on the Relations between Organic Regulations and Cognitive Processes*. Chicago: University of Chicago Press.
Rawls, John. 1971. *A Theory of Justice*. Cambridge, MA: Harvard University Press.
Ruse, Michael. 1986. *Taking Darwin Seriously: A Naturalistic Approach to Philosophy*. Oxford: Blackwell.
Scanlon, T.M. 1998. *What We Owe to Each Other*. Cambridge: Harvard University Press.

Shafer-Landau, Russ. 2012. "Evolutionary Debunking, Moral Realism and Moral Knowledge." *Journal of Ethics and Social Philosophy* 7 (1):1–37.

Singer, Peter. 2011. *The Expanding Circle: Ethics, Evolution, and Moral Progress*. 2nd ed. Princeton: Princeton University Press.

Sober, Elliott, and David Sloan Wilson. 1998. *Unto Others: The Evolution and Psychology of Unselfish Behavior*. Cambridge, MA: Harvard University Press.

Wielenberg, Erik J. 2010. "On the Evolutionary Debunking of Morality." *Ethics* 120 (3):441–464.

Wilkinson, Richard G., and Kate Pickett. 2010. *The Spirit Level: Why Equality Is Better for Everyone*. London: Penguin.

Wilson, David Sloan, and Edward O. Wilson. 2008. "Evolution 'for the Good of the Group'." *American Scientist* 96:380–389.

6 Natural moral values and moral progress

The arc of the moral universe

In a press conference shortly after the death of Nelson Mandela, Barack Obama said of Mandela that "he took history in his hands and bent the arc of the moral universe towards justice." Some individuals, by the force and magnitude of their beliefs and actions, do help to bend history towards greater social justice. But in its original use, the phrase referring to the arc of the moral universe was about the arc itself: it had a bend in it, and the direction of that bend was towards justice. Martin Luther King often used the phrase in this broader sense, and that is how it was originally meant in what is probably its original use, in a sermon by Theodore Parker:

> Look at the facts of the world. You see a continual and progressive triumph of the right. I do not pretend to understand the moral universe, the arc is a long one, my eye reaches but little ways. I cannot calculate the curve and complete the figure by the experience of sight; I can divine it by conscience. But from what I see I am sure it bends towards justice.
>
> (Parker 1853, 84–85)

Parker, an ironic figure in the US abolition movement of the 1850s (Fellman 1974), believed that moral laws were not made by man but discovered through individual conscience and intuition. In addition to his sermonizing widely on the public lecture circuit, Parker was one of the Secret Six financial backers of John Brown and an avid supporter of the American right to bear arms, despairing that while "[o]nce New Englanders had more firelocks than householders. . . . [N]ow the people . . . neither keep nor bear arms" (Fellman 1974, 673). He was also an intemperate racist who thought that inferior, non-white races would lose out in the great social struggle of the survival of the fittest. For Parker, a significant evil of slavery was the ample opportunities it provided for the production of mixed-race children.

Using Parker's phrase, Cohen (1997) examines the historical example of the abolition of American slavery to show how contractualist thinking might be part of explaining progressive social changes. Along with Cohen, we think that contractualism may be an important part of such explanations. But we also think that the human moral sense is likely to play a role as well, a topic we take up in our next chapter, along with natural moral values, the topic of this chapter. EMR is thus closer to Parker's original claim about human intuition, moral realism, and social justice, although for him the truth of moral laws was to be found in God's good will. For EMR, moral truth is ultimately to be found in the natural kinds we have talked about in earlier parts of the book, and while these kinds of good things may be structural features of the universe of evolutionary biology, other features of this same universe may bend social relationships in other less morally salubrious directions. Not all moral change will be moral progress, and we cannot assume that progress will be the inevitable winner.

Moral progress and natural moral values

The Anglo-American abolition of New World slavery nevertheless provides us with a significant historical episode of moral progress. For Britain, colonial slavery and the international slave trade were of significant economic value; for the United States, slavery was an even larger part of the national economy, and the trade in slaves throughout the South was an integral part of the Southern economy. In both countries, moral thought and argument were instrumental in abolishing slavery and the trade in slaves.

In 1869, in his two-volume *History of European Morals*, Irish historian W.E.H. Lecky made the bold (but carefully hedged) claim that the abolitionist crusade in Britain to end the slave trade and then slavery itself in its American territories was "probably . . . among the three or four perfectly virtuous pages comprised in the history of nations" (Lecky 1895, 153). Nearly a century later, Lecky's claim was famously and importantly challenged by Eric Williams (1944) in his book *Capitalism and Slavery*, which argued that rational self-interest or at least the rational self-interest of the capitalist class was behind the abolition of slavery in the Americas. Slave labour was diametrically opposed to wage labour, and the supremacy of wage labour was required to secure the hegemony of the capitalist expansionism that was driving the industrial revolution. Drescher (1977) and Eltis (1987) subsequently provided influential arguments against this claim, and by the turn of our own century, retrospective accounts of the disagreement between Lecky and Williams seem to be largely on the side of Lecky (Davis 2006, 234–249, 2014, 256–290).

We are not historians, so we will need to move carefully in addressing the literature in this area. We are aided in this endeavour, however, by a number of major historical retrospectives on the Anglo-American abolition of slavery that have recently appeared, including but not limited to Blackburn (2013), Davis (2006, 2014), Drescher (2009), Foner (2010), Levine (2013), Oakes (2013), and Reynolds (2005). These books are written by prominent historians of New World slavery for broader audiences, with an aim, in part, to make clear how the perception of slavery as a great and remediable moral evil was key to abolishing both slavery and the slave trade.

For most of these historians, the perception of slavery as a moral evil was tied directly to the actual moral evils intrinsic to slavery as it existed in the British Empire and in the antebellum United States. The reason abolitionists and their supporters thought slavery and the slave trade were evil was in large measure because they were. For philosophers, the argument is not so straightforward. For philosophers, the Anglo-American abolition of slavery example may not appear to offer any genuine evidence that morality starts with moral values rather than moral capacities.

What is clear from the historical literature is that human moral capacities were deeply involved in the abolition of slavery in both cases, and these capacities were tied to very real parts of the material world, such as the pain and humiliation of being whipped, raped, or sold at auction. So, on an immediate reading of the historical literature, a Street-style argument of the kind we considered at the end of Chapter 1 remains possible: we have moral capacities that respond to real aspects of the word, but these real aspects of the world are not themselves moral values, they are just very real things like the physical and emotional sufferings of sentient and reasonable individuals, sufferings of a kind that we find morally disagreeable because of the structure of our moral capacities.

We think that along with moral arguments, human moral capacities are important for explaining the abolition of slavery. But we also think these capacities are ultimately tied to the natural moral kinds discussed earlier in the book. The point of this chapter is to see if we can find an explanatory role for any such values in providing a more complete explanation of this particular episode of moral progress. This is an important part of our argument that the natural kinds at the centre of EMR are moral kinds.

What we are doing in this chapter is similar to what we think de Waal is essentially doing in the broader context of comparative psychology: looking to see how particular kinds of natural moral values might manifest themselves within a particular species' immediate social environment, and then looking to see if members of the species in question are able to recognize and respond to these values in morally appropriate ways. The two values we examine are helping and caring for others. Our discussion here will be as much historical as it is philosophical: ours is an argument for which the details matter.

We should again stress that the argument of this chapter is not meant to be what historians would call whiggish. We do not think that moral progress is an inevitable aspect of human thought or history. If there is an arc to the universe of human morality, there is no guarantee that it will be bent in the direction of greater human justice or well-being. Too many other aspects of our human social and psychological existence affect our relationship to moral values. For creatures as smart as we are, moral values are easy to miss, even when they are staring us in the face. It is arguable that some very important moral values were there to be missed as slavery became deeply embedded in the economies of both Britain and the United States. In particular, had the North not convincingly won the American Civil War, it is not clear when slavery would have ended in the Americas. Elsewhere in the contemporary world, other forms of slavery continue to thrive.

Helping others

In *The American Historical Review* of the 1980s, there is a particularly rich and extended interchange between Thomas Haskell, David Brion Davis, and John Ashworth on the relationship between capitalism and humanitarianism in the Anglo-American abolition of slavery and the slave trade. Davis (1987) and Ashworth (1987) comment on an earlier, two-part paper by Haskell (1985), to which commentaries Haskell (1987) offers a lengthy reply. All three papers are concerned with charting a middle course between Lecky's 1869 claim that ending slavery and the slave trade was a remarkable triumph of humanitarian thought and action and Williams' 1944 Marxist challenge to this claim. The rise of capitalism was important, and it played a role, or more likely roles, in the Anglo-American abolition of slavery, but exactly where and how?

We want to highlight one aspect of Haskell's 1985 argument in a way that he does not. According to Haskell, long before the British abolition movement of the late eighteenth century, people were aware of the evils of slavery. They were indifferent to these evils, regarded them as necessary evils, or viewed them with what Haskell calls passive sympathy. Slavery was unpleasant, but there wasn't anything one could do about it. What changed with the rise of capitalism and the industrial revolution, according to Haskell, was the emerging capacity for individual citizens to harness causal mechanisms made possible by the free market to act collectively, to great effect and at a great distance. This new form of social power moved abolishing national and even international institutions of slavery from outside the bounds of minimally decent Samaritanism to inside its bounds.

To take a well-worn philosophical example from Singer (1972), anyone passing by a small child drowning in shallow water has an obligation to pull the child out of the water, if the person can do so at a comparatively small cost to his or her own life or well-being. This sort of minimally decent Samaritanism is morally obligatory rather than morally supererogatory. But we cross, at some point, the line between obligation and supererogation. If the costs of helping are sufficiently high, there is no obligation to help, even if the threat to those who need help is dire. On Haskell's view of abolition, the possibility of comparatively low-cost action at a distance engendered by the rise of free market capitalism moved helping slaves from outside the bounds of minimally decent Samaritanism to inside its bounds, from supererogation to moral obligation.

Haskell's account of the relationship between capitalism and humanitarianism is philosophically sophisticated, and it is closely tied to arguments over an approach to moral rules that involves drawing moral lines between what has been called an individual moral prerogative to benefit oneself and more general social obligations to benefit others. In addition to Singer (1972), this approach to moral rules is more fully explored in (for example) Thomson (1971), Scheffler (1982), and Unger (1996). The philosophical literature here is extensive and its arguments are contentious. Although some of these controversies are relevant to the argument we are developing here, they do not particularly matter to what we take to be Haskell's key historical insight: because of changing social conditions, the possibility of helping slaves appeared in the social environment of those who became part of the social movement in Britain to abolish first the slave trade and then slavery itself.

Questions about where best to draw moral lines between individual prerogatives and more general obligations to help others are relevant to deeper issues involving the relationship of moral values to moral capacities and finally to wide reflective equilibrium. Individual prerogatives do seem to be a fundamental aspect of human morality, and thus questions over the extent to which they might be patterned by underlying moral values, the structures inherent in the human capacity for morality, or social processes of wide reflective equilibrium raise important issues for the approach to morality we are developing in this book. Nevertheless, such questions can be safely ignored in the context of the argument of this chapter. We focus here on the key insight of Haskell's account, the emergence of the possibility of helping others who were otherwise helpless.

We think that on this central point, Haskell's argument is correct: because of social and economic changes, possibilities for helping slaves appeared in the social environment of the British abolitionists in the late eighteenth and early nineteenth centuries. These changes enabled the British abolitionists to create what would become one of the first social change movements for the reform of social institutions both nationally and internationally. This larger possibility of rendering aid was made up of a myriad of smaller opportunities for participating in or otherwise furthering the ends of this early movement for social change. In the antebellum and wartime United States, a very different range of opportunities for helping captive slaves arose and presented themselves to a very different range of actors in very different social circumstances. What stayed the same was the general appearance of the possibility of helping, which was then noticed and acted upon. This is the same general kind of moral value that is involved, on our view, when a chimpanzee or a small child notices a dropped marker or a rat notices a confined conspecific that can be released from its captivity. The general moral value arises in the right kinds of social environments, where it is subsequently noticed and acted upon. Because acting on moral values has survival value in certain kinds of social environments, moral capacities evolve and influence the development of more sophisticated forms of moral value. From an evolutionary perspective, helping behaviour starts simple and gets more complex.

We will return to these theoretical points at the end of this section. To make them more cogent, we need to explore some of the empirical details of the Anglo-American abolition of slavery. In the British case, and then in the American case, what new possibilities for rendering aid appeared to abolitionists? How were these opportunities and their subsequent exploitation pivotal to the abolition of slavery?

Hochschild (2005) tells the story of the British movement to end the slave trade thoroughly and well (see also Davis (2006, 231–249) and Drescher (2009, 205–241)). In 1784, a 21-year-old divinity student named Thomas Clarkson won an esteemed essay prize at Cambridge University for an essay written in Latin on the horrors of enslaving others against their will. In 1785, he was on a horse headed to London to begin his career. Several times he stopped his horse, increasingly agitated by points from his essay, until finally,

> Coming in sight of Wades Mill in Hertfordshire, I sat down disconsolate on the turf by the roadside and held my horse. Here a thought came to my mind, that if the contents of the Essay were true, it was time some person should see these calamities to their end.
>
> ((Hochschild 2005, 89), quoting from Clarkson (1808, 210))

The question was, how? How could a person help to bring an end to slavery as it then existed? Seeking to publish his essay, in English, Clarkson found himself joining with a group of abolitionists, mainly Quakers, who begin meeting together in 1787 in the London printing shop of James Phillips. In addition to Clarkson, the second central figure in this group was Granville Sharp, who until then had been a writer of largely unsuccessful polemical screeds and a rather more successful defender of individual blacks seeking freedom in court cases against their colonial owners who had brought them into Britain as unpaid servants. One noteworthy case involved a slave named James Somerset. The case was well followed in the press, and although the eventual judgment in Somerset's favour was carefully phrased to apply only to his own case, this judgment was widely interpreted as having outlawed slavery within Britain itself.

At this point, Britain was extensively involved in the Atlantic slave trade and in slavery itself in its remaining American colonies. Drescher (2009, 206) tells us that "[b]y the end of the eighteenth century, British slavers were landing 50,000 slaves per year in the Americas and moving nearly 60 percent of the total number of captives shipped across the ocean." In the meantime, between 1775 and 1807, British-controlled territories went from exporting one third to well over half of all the sugar that reached Europe. Slavery, and the slave trade, represented robust portions of the British economy. The group centred around Clarkson and Sharp chose to focus on the slave trade as the easier of the two evils to challenge. Given the high levels of slave mortality in producing commodities like sugar, stopping the trade would deal a hard blow to colonial slavery itself. The group began publishing and distributing pamphlets, organizing and advertising public meetings, and coordinating petition drives that had an effect on prominent British Parliamentarians (William Wilberforce and William Pitt the Younger) and eventually on Parliament itself, which abolished the British slave trade in 1807.

The pamphlets and public meetings were key factors in igniting broad public interest in ending the slave trade. A widely reproduced and circulated poster gave visual force to the conditions of transport on a slave ship, showing in cross section the slave decks of the ship with slaves packed into them like row after row of sardines in a tin. Pamphlets and talks gave vivid and personal accounts of what this poster meant for those whose plight was portrayed in it. Olaudah Equiano, a freed slave, wrote an important autobiography and lectured widely on his experiences. Also powerful was a book by John Newton, the author of "Amazing

Grace" and a once-respected slave ship captain who later became an equally well-respected clergyman. Through well-distributed publications and well-organized public meetings, personal stories like these found increasingly wider audiences throughout Britain. They vividly brought the ugly details of slavery into what was in the process of becoming the public imagination.

In their different ways, all the individuals involved in this movement for social change saw new opportunities for helping to end slavery as they arose; and seeing such opportunities they seized upon them, from Clarkson and the other members of his group to William Wilberforce to all those British subjects who, in increasing numbers, signed petitions that could not be ignored, even in the face of a well-funded and politically influential lobbying effort mounted in defence of slavery and the slave trade. In 1791, commenting on a lost vote in favour of abolishing the slave trade, a back-bench Parliamentarian commented,

> The leaders [Wilberforce, Pitt, Fox and Burke], it was true, were for the abolition; but the minor orators, the pygmies, would . . . carry this day the question against them. The property of the West Indians [i.e., those with property and investments in the West Indies] was at stake.
>
> (Drescher 2009, 219)

It is important to remember that at this moment of British history, most British subjects could not vote and so would have had very little political voice in the absence of the social movement that resulted in the steady flow of increasingly longer petitions to Parliament. In the end, the proslavery side was right about the economic impact on England of abolishing both the slave trade and slavery: it was significant (Davis 2006, 240–246).

The British slave trade was abolished in 1807. As the British abolitionists continued to push for the abolition of slavery itself, which would not happen in Britain's American colonies until 1833, the abolitionist movement became an international social movement (McDaniel 2013b, 1–18). American abolitionists like William Lloyd Garrison and Frederick Douglass travelled to Britain and Europe to address public meetings and to learn what they could from British abolitionists regarding how to create a successful abolitionist social movement in the United States. From the other side, the Americans were interesting to the British and continental Europeans because of the much greater extent of suffrage in the United States. In terms of overall social equality, the Americans were a far more progressive social experiment in terms of democratic rights than Britain or western Europe. As slavery was abolished elsewhere in the Americas in the first half of the nineteenth century, slavery in the United States became an increasingly peculiar institution, in part and paradoxically because the United States was ahead of Britain and Europe in terms of individual citizens having constitutionally protected political rights to life, liberty, and the pursuit of happiness.

Despite these democratic advantages, there were other significant aspects of the US situation that made abolitionism in the United States impossible as a broadly based social movement. US abolitionists like Garrison and Douglass remained a small minority, widely reviled and sometimes open to physical assault. In the United States, the slave economy was deeply embedded in the economy of the South and of the United States as a whole. Drescher (2009) sums up the situation neatly:

> The sheer magnitude of the institution of Slavery in America [the US] had always been the most formidable barrier to envisioning any practical, peaceful means to its rapid end. . . . [I]n the South, Slaves had become the region's major source of wealth after the value of the land itself. . . . [In 1860] the gross national product of the entire United States was only about 20 percent more than the value of its southern slaves, equivalent, in today's terms, to nearly $10.5 trillion dollars.
>
> (296)

> By 1840, America [the US] provided more than 60 percent of the Atlantic world's cotton, a proportion that rose to more than 80 percent by 1860.
>
> (297)

> Between the 1820s and 1860, the cotton South provided about half the value of US exports within the United States, and the antebellum South grew faster per capita than the north between 1840 and 1860. . . . [W]ith the world's third highest per capita income, it [the South] ranked above France, the Germanies, or any other region with more than ten million or more inhabitants.
>
> (298)

Formidable though the economic barrier was, the problems in the United States went well beyond this single large obstacle. Unlike any group in Britain, Southerners were engaged in a long running argument that slavery was a legitimate moral good and not just for economic reasons. First, slavery was fully justified by the Bible, in ways outlined by Davis (2006):

> In the old testament God tells Moses that the ancient Israelites should take their male and female slaves "from the nations around you . . . [t]hese shall become your property: you may keep them as a possession for your children after you, for them to inherit as property for all time."
>
> (187)

> [Since] most Southern Christians fervently believed in the descent of all humans from Adam and Eve . . . Ham's sinful contempt for his father provided a way of distinguishing the animal-like "Canaanite race" from the superior descendants of the "fair and comely" Japheth.
>
> (187)

> The curse of slavery was even good for the Canaanite race, since . . . "the excesses of his animality are kept in restraint and he is compelled to live an industrious, sober life, and certainly a more happy one than if he was left to the free indulgences of his indolent savage nature."
>
> (187)

Combined with this later argument was an argument against the conditions of wage labour in the north. Slaves were employed for life, and well taken care of as valuable pieces of property. They were not expendable pieces of living machinery to be used up and tossed out. And better black slaves should live an industrious and Christian life on quiet plantations in the US South than face the rigors of savage life in Africa, whence their wild brethren had hunted them down and sold them into slavery. (For contemporary examples of these sorts of arguments, see Mary Eastman's (1852) slave-owner's response to the now much more famous *Uncle Tom's Cabin*.)

As this last point would suggest, racism ran as a deep current through both the antebellum northern and southern states. Some northern states and territories, for example, took measures to prevent free blacks from settling within their borders. So, not only would emancipated slaves somehow have to be fairly paid for in compensation for each individual slave-owner's substantial loss of privately held property, but something would also have to be done with all the newly created wage labourers such a system of nationwide emancipation would suddenly create. Because blacks were widely considered to be inferior to whites and most whites were prejudiced against them, blacks would not be able to fairly compete as wage labourers; and to the extent that they could compete, such competition would drive down wages for all workers. For those who had their doubts about the literal truth of the Biblical story of Adam and Eve, the early nineteenth century also saw the advent of scientific racism and the polygenic idea that the human race did not include all *Homo Sapiens* (Lowance 2003, 249–326). Blacks were considered an intermediate race between apes and whites, and they thus were cursed by religion, science, or both. Racist views were also held by many abolitionists, some of them as radical in their abolitionist views as Theodore Parker (Lowance 2003, 299–310).

For all these reasons, the most sensible US antislavery position before the Civil War was probably that of Abraham Lincoln: gradualism and some form of voluntary colonization (Foner 2010, xx–xxi, Oakes 2013, xii). Because it would be far too costly to compensate owners for the immediate emancipation of their slaves, slavery would have to die a slow and natural economic death (which in Lincoln's own estimation would probably take about a hundred years), and because few whites would freely want to labour along with blacks and blacks themselves would not want to be met by such hostility or by their perceived inability to compete on a level playing field with members of an allegedly superior race, the best option for slaves who gradually found themselves freed would be the voluntary colonization of various parts of Africa or South America. In the decades leading up to the Civil War, some wealthy and influential black leaders had indeed explored this option, including James Forten, who was an important financial supporter of William Lloyd Garrison. Davis (2014) charts at length the history of early nineteenth-century colonization initiatives in the United States. It was not, as one might expect, a satisfactory option. In 1831, Garrison began publication of *The Liberator*, adopting a radical abolitionist position of immediate emancipation.

Here it is important to note that Lincoln's gradualism was not a passive antislavery position. Lincoln and the Republicans were widely regarded in the South as abolitionists, if not radical abolitionists. A key element of the Republican platform was confining slavery to the states where it then existed. With the slave states and their economies increasingly dwarfed politically and economically by a fast-growing rest of the United States, the peculiar institution of slavery would inevitably wither and die. That this was a very real threat was evidenced by the secession of the southern states that left the Union to form the Confederate States of America after Lincoln's election. When the Confederate South fired on Fort Sumter, the Union went to war against what it took to be illegal acts of secession by disloyal citizens in the states involved in the formation of the Confederacy.

Against this backdrop and important for our argument, the resulting Civil War of 1861–1865 created possibilities for immediate emancipation that would have been unthinkable before the war, and, indeed, they only became thinkable because of the actual course of the war. If freeing captive slaves from their bondage is an important form of helping those who can be helped, on numerous different occasions during the war this value appeared in sudden and unanticipated ways. Appearing when and where it did, this value and the moral responses to it did much to change the course of the war. On our interpretation of these events, moral values such as helping those who can be helped existed as very real parts of the causal landscape of the human world.

At the outset of the war, Lincoln was bound by the US Constitution to respect the positive laws of the states that had created property rights in slaves (Foner 2010, 42–43 and 165; Oakes 2013, 1–22). During the war, some slave states remained loyal to the Union as did some slave owners in those states where newly created governments joined together with each other to form the Confederacy. Lincoln regarded this last move as illegal but nonetheless took himself to be bound by the Constitution to respect property rights in slaves where such rights otherwise existed. At the outset of the war, the war could only be over secession, not over slavery itself. This political situation changed rapidly during the course of the war.

We pick three pivotal points where an opportunity arose to help free captive slaves and was seized upon by the actors involved. The first case, described in detail in Oakes (2013, 90–105), occurred near the outset of the war and involved a Union General, Benjamin Butler, who had been posted to Fortress Monroe in Virginia in 1861. On the evening of May 2, a day after he arrived at Fortress Monroe, three escaped slaves crossed over to Butler's lines. A Confederate officer appeared the next day to reclaim the slaves for their owner, based on the Constitution's protection of private property in slaves. While outraged by the secessionists, Butler had been a proslavery Democrat before the war. On the other hand, slaves were busily engaged in the construction of nearby gun batteries that probably couldn't be constructed without their labour. Acting on impulse, Butler declared that the Confederate officer was in no position to claim any Constitutional rights, given that he must consider himself, as a secessionist, to be an officer of a foreign army. Butler kept the slaves but issued receipts for them that could be redeemed after the war by their owner.

Butler immediately wrote to his superior officers to validate what he had done. This set off a flurry of activity that extended to Lincoln and to Congress, generating sets of military instructions and several Acts of Confiscation. None of this activity legally emancipated any slaves; instead, runaway slaves became increasingly referred to as contraband of war. As the war continued and slaves continued to cross over to Union lines and were offered employment in support of Union troops, it became clear that they were trustworthy and loyal. They could be depended upon for correct information about Confederate troop movements and fortifications or to support Union troops in a variety of ways. This created the possibility that runaway slaves could themselves be part of the North's winning the war, as autonomous persons rather than as confiscated property.

While Butler himself, in the initial incident of May 1861, may not have set out to help the slaves who had escaped to his lines, other Union officers did see themselves as doing this, and the slaves certainly saw matters in this way. Escaped slaves who returned to their owners suffered dire fates. As the war continued, escaping to Union lines increasingly became a path to freedom. If you appeared and needed help, you were likely to get it.

As a second case of new possibilities of help arising and being quickly recognized as such, we consider the Emancipation Proclamation of 1863, discussed in detail in Foner (2010, 206–257). This was a war measures act and as such an executive act of the President as commander in chief of the military. It did not end slavery. It proclaimed that certain slaves in certain areas of the Confederacy were forever free. Whether they would remain so was not clear. This would depend upon how exactly the war ended, and how courts dealt with slavery under the US Constitution as it was then written.

Our point here is that the Proclamation was a war measure: it was a way of helping some slaves in some parts of the South that arose as a necessary part of pursuing the war at a vital junction in that war. The Proclamation threatened the South as a whole with deep social unrest. The Proclamation offered immediate emancipation, without compensation for slave owners, and it did not distinguish between loyal or disloyal owners. It was a way of helping to free slaves that arose as it did, when it did. And although it was motivated by concerns that went beyond those of simply freeing slaves, it was also motivated, in part, by the idea that the time had now come to end slavery sooner rather than later. A new attitude towards free slaves enabled Lincoln to include the final measure of the Proclamation: slaves could enlist as Union soldiers. This they did, in sizeable numbers, distinguishing themselves in battle as loyal Union soldiers. Gradualism and colonization were dead, and freed slaves were on their way to becoming citizens. They were now fully part of the war effort to save the Union.

The Thirteenth Amendment to the US Constitution was the final act of helping US slaves become free. Such an amendment became thinkable only because of the way in which the war unfolded. Lincoln and the Republican party had been antislavery from the beginning of the war, but it was only the war that enabled them to end slavery when and as they did, in a way that they themselves would never have considered possible upon Lincoln's election in 1860. The war was long and bitter, and slavery was now widely recognized to be its cause. How best to guarantee lasting peace? Oakes (2013, 430–488) provides a telling account of the process that led to the ratification of the Thirteenth Amendment that finally brought to an end the Constitutional impasse over slavery. Slavery would no longer be a matter of the rights of individual states: it would be federally abolished. To rejoin the Union, formerly Confederate states would have to ratify the amendment, which they did. The Thirteenth Amendment was one of several ways Lincoln and the Republicans in Congress contemplated ending slavery once and for all in the United States. For practical reasons, it emerged as the best alternative.

In each of the many steps towards complete emancipation, helping actions became possible that could not have been foreseen. Particular opportunities to help captive human beings escape the bonds of slavery appeared in a wartime environment where they were noticed and acted upon. The motives may have been mixed, but helping slaves out of the bonds of slavery was almost always among them. If helping another is a natural kind, we humans seem to respond to this kind in a moral sort of way. The natural kind of helping another thus appears to be a moral kind, one that arises across the environments of a wide range of social and intelligent creatures. In each of these environments, it is realized in different ways, but it seems to be the same general kind of thing, whether it is an ape passing back a marker that has been dropped or a Union officer giving someone his or her freedom by passing back a receipt for war contraband.

If EMR is correct, morality and moral normativity appear in the biological world in more or less complex sorts of ways. Initially, natural moral values have normative power insofar as they are naturally attractive or repellent. This normative power is neither wholly in the object nor in the response: it is instead in the evolving biological circuit that includes both the value and the response. Such circuits create the original form of moral normativity. The increasingly complex psychological capacities that develop from such initial responses become themselves further sources of moral normativity. Along our own moral trajectory, the normative power of morality culminates with language, thought, and the human capacities for reason and rationality that enable us to reach stable points of wide reflective equilibrium. Natural moral values may thus be expected to be an important part of explaining historical episodes of moral progress.

Caring for others

In the opening pages of *Regarding the Pain of Others*, Sontag (2003) raises the point that when we look at photographs of bodies torn apart by war, it matters who the "we" is who is doing the looking. The photographs do not immediately speak for themselves. If the bodies belong to our group, or a group we are close to, we are immediately drawn into the world disclosed to us by the camera. On the other hand, if the bodies in the photograph are those of our enemies, our empathy will not be so quickly engaged. War is hell, and these things happen. It is too bad that this is so, but this kind of suffering has to be put into its broader perspective of a necessary or just war, or if nothing else, an unstoppable war.

Several months after the publication in 1845 of *Narrative of the Life of Frederick Douglass, an American Slave*, Douglass was on his way to lecture in Britain, attracted by the reported fact that British audiences were amazingly free of racial prejudice and also by the fact that in Britain he would be safe from recapture as a fugitive slave, especially because his owners, Hugh and Thomas Auld, were incensed over the way they had been portrayed in the *Narrative* (Davis 2014, 291–292). While Douglass had achieved great success in the northern United States as a speaker, and his book had become a bestseller, the abolitionist position he represented was widely reviled in the United States. The painful details of enslavement, while vital to galvanizing support for the British social movement to end the slave trade and slavery in its American territories, were not having the same effect in the United States. For the reasons outlined earlier, abolitionists like Douglass and Garrison were unable to find the same traction in the United States that the antislavery movement in Britain had found among the general population.

Perhaps this is part of the broader explanation of the break between Douglass and Garrison, which was more immediately explained by Douglass in his second book, more tellingly and personally titled *My Bondage and My Freedom*:

> During the first three or four months, my speeches were almost exclusively made up of narrations of my own personal experience as a slave. "Let us have the facts," said the people. So also said Friend George Foster, who always wished to pin me down to my simple narrative. "Give us the facts," said Collins, "we will take care of the philosophy." Just here arose some embarrassment. It was impossible for me to repeat the same old story month after month, and to keep up my interest in it. It was new to the people, it is true, but it was an old story to me; and to go through with it night after night, was a task altogether too mechanical for my nature. "Tell your story, Frederick," would whisper my then revered friend, William Lloyd Garrison, as I stepped upon the platform. I could not always obey, for I was now reading and thinking. New views of the subject were presented to my mind. It did not entirely satisfy me to *narrate* wrongs; I felt like *denouncing* them.
>
> (Douglass 1855, 361)

But as Sontag (2003, 26–27) points out, photographs of atrocities need to be crude statements of fact to move us as fully as we might be moved. Staged photographs or photographs that look too artistic fail to impress us in the way that unstaged pictures do: "by flying low, artistically speaking, such pictures are thought to be less manipulative" (27).

In light of this point, we might also consider who was looking at, or in this case listening to, the atrocities being depicted at the public events where Douglass would have been speaking. For many listeners, slaves would have been the literal descendants of Ham, justifiably cursed. And on the whole, slaves may have been better off on plantations than otherwise: "[t]he Negro, when a 'slave' to a Caucasian, is vastly higher in the scale of humanity, than when in his native state . . . because our heavenly Father made him an inferior being, a perpetual child" (Reynolds 2005, 115). Slaves were natural subordinates, and so the subordination of their autonomy to that of another was not a great moral evil. While they sometimes might suffer under the bonds of slavery, such suffering had to be put into its broader context.

One can thus see a point to Garrison's insistence in getting Douglass's own particular sufferings into the harsh bright light of the auditory equivalent of an unstaged photograph. On the other hand, one can equally understand Douglass's frustration at being treated like a walking photograph, to be pulled out, shown around and discussed like an object. As he himself made clear, Douglass certainly felt like he was not fully respected on an intellectual level by Garrison and other Garrisonians. Yet Garrison's verbal attacks on Douglass after their split were just as pugnacious as his attacks on others he had broken with in the movement, which suggests some measure of equal respect on an intellectual level (McDaniel 2013a).

The deeper problem involved in the ideological splits between Garrison and other Garrisonians, including Douglass, was no doubt the floundering nature of the abolition movement in the United States. The great abolitionist movement for social change was not working in the United States as it had worked in Britain. In addition to the factors mentioned in the preceding section, part of this difference may have been a more general problem with empathy. Sontag (2003, 99) considers a woman in Sarajevo who bitterly recounted that when her television was showing pictures of the destruction of Vukovar, she turned it off. So, how could she then criticize those outside the former Yugoslavia, who also turned their televisions off? Sontag's response is that empathy is often tied to the ability to do something – if we are helpless, we turn our empathy off.

In the antebellum United States, John Brown was someone who could not turn his empathy off and also someone who could not stand idly by while slaves were being violently assaulted. Indeed, Brown saw the whole institution of American slavery as one long and continuous violent assault against those who were enslaved by it. Given the extent to which this violence was tacitly accepted throughout the United States, the only way to successfully challenge it was by a spectacular act of terrorism. This is what Brown meant the raid on Harpers Ferry to be, and this is how the raid was interpreted by the South, which saw it as what we might anachronistically label its 9/11 wake-up call to arms.

As detailed in Reynolds (2005), Brown was an exceptional individual. His empathy was engaged with regard to slavery by his strict Calvinist upbringing and through his early childhood friendship with a black slave boy of his own age. While the other boy displayed the same moral and intellectual virtues as Brown, Brown was praised while the other boy was consistently abused. The young Brown felt the position of children like his friend was entirely wretched and hopeless, with "neither Fathers nor Mothers to protect and provide for them." Brown thought all children deserved equal respect, black or white, male or female, a moral view he impressed upon all of his children, several of whom received mortal but not immediately fatal wounds during the protracted fighting at Harpers Ferry. Brown himself was run through with a sword and spent most of his trial on a cot. In early meetings with him, Frederick Douglass was particularly impressed with Brown's empathy, not just because of how Brown treated him personally but because Brown was fully prepared to die to set slaves free. Although Douglass had declined Brown's invitation to come with him to Harpers Ferry, in a speech in 1881 Douglass declared that Brown could be called "our noblest American hero" (Reynolds 2005, 492).

As an adult, Brown knew and worked with blacks. He respected them as equals and they him. Some of them followed him to the gallows in the aftermath of the raid, which was planned to be the beginning of a protracted guerrilla war against the South. Munitions would be seized from the armoury at Harpers Ferry, and the small vanguard group led by Brown would flee to the nearby Blue Ridge Mountains. There they would be joined by increasing numbers of escaping slaves in what would grow to become the largest slave rebellion in North America. Brown believed that given the opportunity, blacks would join the rebellion to fight for their freedom, and blacks and whites would work together in the mountains, guided by a model Constitution Brown had in his possession when he was captured. The model Constitution treated all people equally, including blacks and women, and put blacks into leadership roles over whites (Reynolds 2005, 249–255). Garrison was a radical abolitionist in calling for the immediate abolition of slavery; Brown was far more radical, advocating not only for the immediate and violent overthrow of slavery but with its abolition full and equal social and political rights for all freed slaves. With the raid on Harpers Ferry, Brown knew where he was aiming. Slave rebellions were the great fear of the South. While Southern elites claimed plantation life was benign and most slaves loyal to their masters, any hints of rebellion were brutally quashed, with the successful slave rebellion in Haiti held up as a fearsome and real possibility. As a freed slave, Douglass saw the raid on Harpers Ferry as suicidal, and he was proven right.

Lincoln was hardly an abolitionist at the beginning of the Civil War, but the rapid evolution in his own thinking about blacks over the course of the war made him almost as exceptional as Brown in terms of empathy. Although he was born and grew up in slave states, Lincoln himself had very little personal contact with slaves. He thought that slavery was wrong, but that given their physical differences, blacks and whites could never live together as fellow citizens. In 1864, Lincoln wrote,

> I am naturally anti-slavery. If slavery is not wrong, nothing is wrong. I can not remember when I did not so think, and feel.
>
> (Foner 2010, 3)

But he also wrote

> If all earthly power were given me, I should never know what to do. . . . My first impulse would be to free all the slaves, and send them to Liberia, – to their own native land. But a moment's reflection would convince me . . . the sudden execution [of such a plan] is impossible. . . . What then? Free them and keep them among us as underlings? Is it quite certain that this betters their condition? . . . Free them, and make them politically and social, our

> equals? My own feelings will not admit of this; and if mine would, we well know that those of the great mass of white people will not.
>
> (Foner 2010, 67)

With the war underway and slaves crossing over to Union lines, Lincoln no longer had time for such casual reflections. In August 1862, he summoned a committee of black leaders to the White House – an unprecedented event – to push them on several colonization initiatives his administration was then actively exploring. While he understood their great suffering under the burden of slavery, he thought it should be clear to them that they would never live in a socially equal relationship with white Americans, and that they would bear some of the blame for not pursuing more promising alternatives to their increasingly problematic presence in America (Davis 2006, 318–319).

As the war progressed, Lincoln's attitudes towards blacks rapidly changed. In part this was because of his continuing meetings with black leaders, who, like Douglass, had well-developed views about what we would now call race relations in the United States. When Douglass, a sharp critic of Lincoln's policies, met with him at the White House in 1863, he recalled being impressed by Lincoln's "willingness to engage in discussion without ever remind[ing] me of the . . . difference in color" (Foner 2010, 256). But the other great trigger of empathy towards blacks, for Lincoln and for the North more generally, was the performance of black soldiers. The valiant exploits of black troops were widely reported on, and these popular reports helped demolish ideas of blacks' childlike docility or savage barbarism (Foner 2010, 256–257). In his second inaugural address of 1864, Lincoln in effect vindicated Brown's view that the inherent violence of slavery was a national sin and that the wages of that sin were the many deaths of the war:

> Fondly do we hope – fervently do we pray – that this mighty scourge of war may speedily pass away. Yet, if God wills that it continue, until all the wealth piled by the bond-man's two hundred and fifty years of unrequited toil shall be sunk, and until every drop of blood drawn with the lash, shall be paid by another drawn with the sword, as was said three thousand years ago, so still it must be said "the judgments of the Lord, are true and righteous altogether."
>
> (Foner 2010, 325)

As Brown had argued, slaves were worth dying for. They were active participants in America's antebellum achievement of great wealth and productivity, and they were not to be sent away to colonies in Africa or South America. In the United States, the era of slave children with no parents to protect them was soon to be over.

Empathy is central to our moral capacities and in moving us to act. But it is also a powerful moral value in and of itself. Together with its effects, it may take time to emerge in a situation. The situation may have to be morally right in other respects. With Brown, we see the early and powerful emergence of empathy as one moral value among others in the larger social context of the abolition of slavery in the United States. In Britain, the situation regarding empathy had been easier: Clarkson may have been agitated when he dismounted from his horse, but his despair was over not knowing how to proceed against a gigantic social evil as a single individual. His empathy was easily aroused by the facts recounted in his essay, and this same level of empathy could easily be aroused in others, if only the technological means to do so could be found.

In the United States, Brown's level of empathy stuck out as something truly remarkable, something recognized by blacks and whites alike. After Brown was hanged for his raid on Harpers Ferry, a black eulogist commented that he "fully, really and *actually* believed in the equality and brotherhood of man. . . . [He] admired Nat. Turner as well as George Washington" (Reynolds 2005, 408). Black churches in Detroit prepared for a month of mourning. Services were held across Haiti, where Brown was proclaimed a martyr. Victor Hugo wrote that "the murder of Brown . . . would penetrate the Union with a secret fissure, which would, in the end, tear it asunder" (Reynolds 2005, 409). And indeed, with his death and from the calm and collected way that he endured his incarceration and trial, Brown quickly became a potent symbol of abolition in both the North and the South. If he had acted violently, slavery was itself a far greater and ongoing form of violence against generations of blacks; once the magnitude of this injustice was truly perceived, it would also be seen that he had acted out of necessity in violently opposing it. If he had to die to prove this point, he was fully prepared to do so.

In the weeks immediately after the raid, Ralph Waldo Emerson, the philosophical conscience of New England, if not the North as a whole, compared Brown's willingness to sacrifice his life for the freedom of slaves and for the national sin of slavery itself to the willingness of Christ to sacrifice himself for the sins of all mankind. Like "the shot heard round the world," Emerson's response to Lexington, a similar highly symbolic shot was fired off when he said of Brown before his hanging, "who, if he shall suffer, will make the gallows glorious like the cross" (Reynolds 2005, 366).

The Southern reaction to this verbal shot is easy to understand: imagine Osama bin Laden being similarly praised by prominent philosophers at Harvard or NYU in the immediate aftermath of 9/11. John Brown's level of empathy, and responses to it in both the North and the South, played an important causal role in the secession of the South and in the war that followed. Union troops marched into battle singing *John Brown's Body*. As the war continued, new lyrics were added to the same tune, and it became *The Battle Hymn of the Republic*: "as He died to make men holy, let us die to make men free."

Abolition and social contracts

Contractarianism does not fare well as an explanation of the Anglo-American abolition of slavery. Abolishing slavery was not in the economic interests of either the British or the Americans, and greater economic opportunities were not what abolitionists were aiming at. To the extent that rational self-interest was involved, it hardly seems to play a dominant role in the explanation of this salient episode of human moral progress.

To return to the argument of Gauthier (2013) that it is rational to value other rational cooperators as potential contractors, there is little evidence that this was much of a motivation for abolitionists in either Britain or the United States. Americans by and large did not view slaves as fully rational or as potentially rational contractors in an expanded social contract that would bring greater benefits to all. It was thought that freed slaves were and would continue to be a social problem, best dealt with gradually and through colonization. The war changed this view on the basis of respect for the courage and loyalty of black soldiers. Even so, attitudes did not change dramatically or completely: blacks in America faced about another hundred years of social and economic segregation after the war that freed all those who had been slaves. Free blacks were not welcomed with open arms into a racially expanded view of the social contract. In his gradualist and colonizing views, Lincoln had not misread social attitudes among nineteenth-century whites in the United States.

Turning to the British abolition movement, ending the slave trade was unlikely to aid Britain economically or to create a class of new and potentially rational contractors. It would just leave more blacks in Africa and make it harder to work to death slaves in Britain's American territories. Although Adam Smith had argued that wage labour was economically superior to slave labour, the most persuasive arguments against the abolition of slavery argued that this was not so, and in this argument based on their own rational self-interest, the British defenders of slavery were proven right. Britain experienced large and sustained economic losses in abolishing first the slave trade and finally in 1833 slavery itself in its colonies. The dramatic economic failures of emancipation in the British colonies and in Haiti were prominently used by US Southerners as an argument against the abolition of slavery in the United States. Contractarians might resort to some sort of false consciousness argument to overcome such obstacles, but that would simply make them defenders of Eric Williams' failed Marxist argument. The main explanatory problem for the contractarian is that rational self-interest was clearly and heavily on the side of those who were defending slavery in both Britain and the United States.

Contractualist theories fare only somewhat better in explaining the abolition of slavery in Britain and the United States. Again, neither the British nor the Americans saw themselves as entering into a broader social contract with freed slaves. While the British may have been more willing to treat blacks as their social and political equals, most of the slaves they freed were in their colonies and not in Britain itself. And while the Americans passed the Fourteenth and Fifteenth Amendments shortly after the passage of the Thirteenth Amendment, they did not think before the war that blacks were their social and political equals, and to a large extent they did not think this after the war. Despite the Fourteenth and Fifteenth Amendments, the blacks freed by the Thirteenth Amendment became socially, economically, and politically segregated from the larger white population. Few white Americans, other than John Brown, would have supposed that should a Veil of Ignorance be suddenly lifted, they might find themselves a black person. It was wrong to enslave blacks, and once granted, their freedom needed to be protected; but they were hardly to be considered the moral equals of whites in the sense of being equal rational contractors in an Original Position of complete social and political equality. Although the Fourteenth Amendment protected the civil rights of blacks and the Fifteenth Amendment gave them voting rights, such constitutionally created rights did not count for very much socially or politically in the period of reconstruction after the war. One might argue that it was not until the civil rights movement and *Brown v. Board of Education* in 1954 that the civil rights of blacks became a real source of social change in the United States.

In examining the abolition of slavery, we have argued that natural moral values seem to be interesting elements in the causal explanation for this significant shift in Western moral thinking. Other moral factors, such as reason and an underlying sense of justice or fairness, no doubt played their own significant roles in bringing about this progressive moral change. We are not arguing that these other factors played no role in what happened, but that these roles were likely augmented by the role played by natural moral values. The natural kinds we discussed in earlier chapters would thus seem to be morally important kinds of things in our human environment. This strongly suggests that these natural kinds are moral kinds.

Bibliography

Ashworth, John. 1987. "The Relationship between Capitalism and Humanitarianism." *The American Historical Review* 92 (4):813–828.

Blackburn, Robin. 2013. *The American Crucible: Slavery, Emancipation and Human Rights*. London: Verso.

Clarkson, Thomas. 1808. *The History of the Rise, Progress, and Accomplishment of the Abolition of the African Slave-Trade by the British Parliament.* London: Longman, Hurst, Rees, and Orme.

Cohen, Joshua. 1997. "The Arc of the Moral Universe." *Philosophy and Public Affairs* 26 (2):91–134.

Davis, David Brion. 1987. "Reflections on Abolitionism and Ideological Hegemony." *The American Historical Review* 92 (4):797–812.

Davis, David Brion. 2006. *Inhuman Bondage: The Rise and Fall of Slavery in the New World.* Oxford: Oxford University Press.

Davis, David Brion. 2014. *The Problem of Slavery in the Age of Emancipation.* New York: Knopf.

Douglass, Frederick. 1855. "My Bondage and My Freedom." http://docsouth.unc.edu/neh/douglass55/douglass55.html.

Drescher, Seymour. 1977. *Econocide: British Slavery in the Era of Abolition.* Chapel Hill: University of North Carolina Press.

Drescher, Seymour. 2009. *Abolition: A History of Slavery and Antislavery.* Cambridge: Cambridge University Press.

Eastman, Mary. 1852. *Aunt Phillis's Cabin: Southern Life As It Is.* Philadelphia: Lippincott, Grambo & Co.

Eltis, David. 1987. *Economic Growth and the Ending of the Transatlantic Slave Trade.* Oxford: Oxford University Press.

Fellman, Michael. 1974. "Theodore Parker and the Abolitionist Role in the 1850s." *The Journal of American History* 61 (3):666–684.

Foner, Eric. 2010. *The Fiery Trial: Abraham Lincoln and American Slavery*. New York: Norton.

Gauthier, David. 2013. "Twenty-Five On." *Ethics* 123 (4): 601–624.

Haskell, Thomas L. 1985. "Capitalism and the Origins of the Humanitarian Sensibility." *The American Historical Review* 90:339–361, 457–566.

Haskell, Thomas L. 1987. "Convention and Hegemonic Interest in the Debate over Antislavery: A Reply to Davis and Ashworth." *The American Historical Review* 92 (4):829–878.

Hochschild, Adam. 2005. *Bury the Chains: Prophets and Rebels in the Fight to Free an Empire's Slaves*. Boston: Houghton Mifflin.

Lecky, W.E.H. 1895. *History of European Morals: From Augustus to Charlemagne*. 3rd ed. 2 vols. Vol. 1. New York: D. Appleton and Company.

Levine, Bruce. 2013. *The Fall of the House of Dixie: The Civil War and the Social Revolution that Transformed the South*. New York: Random House.

Lowance, Mason I., Jr., ed. 2003. *A House Divided: The Antebellum Slave Debates in America 1776–1865*. Princeton: Princeton University Press.

McDaniel, Caleb. 2013a. "The Lives of Frederick Douglass." http://wcm1.web.rice.edu/lives-of-frederick-douglass.html.

McDaniel, Caleb. 2013b. *The Problem of Democracy in the Age of Slavery: Garrisonian Abolitionists and Transatlantic Reform*. Baton Rouge: Louisiana State University Press.

Oakes, James. 2013. *Freedom National: The Destruction of Slavery in the United States, 1861–1865*. New York: Norton.

Parker, Theodore. 1853. *The Sermons of Religion*. Boston: Nichols and Company.

Reynolds, David S. 2005. *John Brown, Abolitionist: The Man Who Killed Slavery, Sparked the Civil War, and Seeded Civil Rights*. New York: Knopf.

Scheffler, Samuel. 1982. *The Rejection of Consequentialism*. Oxford: Clarendon Press.

Singer, Peter. 1972. "Famine, Affluence, and Morality." *Philosophy and Public Affairs* 1 (3):229–243.

Sontag, Susan. 2003. *Regarding the Pain of Others*. New York: Farrar, Straus and Giroux.

Thomson, Judith Jarvis. 1971. "A Defense of Abortion." *Philosophy and Public Affairs* 1 (1):47–66.

Unger, Peter. 1996. *Living High and Letting Die: Our Illusion of Innocence*. Oxford: Oxford University Press.

Williams, Eric. 1944. *Capitalism and Slavery*. Richmond, VA: University of North Carolina Press.

7 Partial and impartial moral reasons[1]

Towards a theory of moral competence

From the point of view of standard approaches to evolutionary psychology and sociobiology, partial moral reasons may seem eminently reasonable: kin selection and reciprocal altruism should dispose us to act kindly towards particular others in particular social relationships with us. More impartial moral reasons will seem to be more problematic: promulgated by the moral idealists among us, like John Brown from our last chapter, such reasons may ultimately trade on nothing more than a human psychological tendency to be taken in by them. Why treat the interests of distant and socially unrelated individuals or groups on a par with the interests of closer and more socially related individuals or groups? Why treat animal interests on a par with human interests? We might worry that even if evolutionary ethics does not reduce down to some form of individual or genetic selfishness, it does at least make an attraction towards various forms of tribalism a particularly deep part of human nature.

This point raises a deeper worry from the point of view of moral philosophy. Whatever psychological limits standard approaches to evolution and ethics might put on human nature, these approaches do not offer us particularly well-developed explanations for how we are to move from moral values or moral instincts to moral reasoning and moral justification. Humans often reason about morality in ways that seem unrelated to our biological fitness, and some moral reasons seem much better than others. So far neither sociobiology nor evolutionary psychology has been able to tell us anything profoundly interesting or important about the patterns of moral reasoning or justification that really matter to us in our contemporary moral world. When it comes to resolving contemporary moral debates, evolutionary ethics simply has not had much in the way of carefully worked-out ethical arguments to offer. Maybe all forms of ethical argument and agreement ultimately turn out to be illusory or a disguised form of self-interest, but maybe not. Ethical discourse certainly seems to involve a good deal of argument and agreement that is not grounded in any ethically interesting ways in individual or genetic self-interest.

In the last chapter, we explored several important ways in which natural moral values may have played a causal role in the important shift in Western thinking about ethics that moved New World slaves from the class of property to the class of self-owning autonomous persons. In this chapter, we shift our attention away from natural moral values themselves to the human capacity for recognizing such values, and in so doing, we explore one important way in which a biologically based human capacity for morality might be causally related to moral reasoning and justification. In particular, we examine how such a capacity might affect ongoing processes of wide reflective equilibrium, as this important concept of moral reasoning and justification is understood by contemporary moral philosophers. Regardless of how well-developed human moral instincts might be relative to those of other primates or mammals, with the distinctively human development of language and thought, our biologically based moral capacities cannot often be expected to enable us to recognize right or wrong in any direct or immediate sort of way. With language and thought, social arrangements become highly complex and variable, and if we recognize them as fair, for example, this recognition will likely be mediated by the social meanings of such arrangements. Even so, what we might think of as our underlying and biologically based moral competence may still play an important causal role in processes of wide reflective equilibrium, which processes will themselves play an important role in the construction of social meanings as they pertain to moral values like fairness.

Processes of wide reflective equilibrium are social and historical. While they involve moral argument and moral reasoning, they are also affected by cognitive and motivational variables that, on our view, are only tangentially related to morality, such as religious belief, self-interest, and hegemonic class interests of one kind or another. Here we set aside such variables as much as we can and as interesting as they might be to focus instead on the idea that processes of wide reflective equilibrium both provide and rest upon provisionally fixed points of moral agreement, what John Rawls (1971) and Norman Daniels (1979, 1980) have called our considered moral judgments. We agree that such judgments are largely fixed through a coalescing and solidification of convincing moral arguments; however, we also think that the cogency of many of these arguments may ultimately depend in some way or another on an underlying human moral competence with its own internal structures. While we are not in a position to work out a fully developed theory of human moral competence, we focus on what we take to be two important aspects of such a theory: impartiality and moral forms of partiality.

Heterodox moral perceptions

We begin with an example of a contemporary moral issue, the issue of whether there are any such things as universal or transnational human rights. From the point of view of convincing arguments and the wide equilibria they are part of, heterodox moral perceptions – forms of moral response not inculcated by social learning or maintained and reinforced by social sanctions or symbolic practices – are interesting because of how they upset moral equilibria and set them off in interestingly new moral directions. For EMR, such perceptions are also interesting because they seem to be relatively direct and immediate.

Appealing to an argument much like the one in Cohen (1997) regarding the obvious evils of slavery, Drydyk (1997) defends the idea of transnational human rights against the claim that such rights are ineluctably Eurocentric and hence biased by an invidious form of moral partiality. On Drydyk's view, there are certain bad things that can happen to human beings, things that can be understood to be bad not just by those to whom they happen but by anyone willing to engage in free and uncoerced dialogue with those to whom they happen or threaten to happen. In such a dialogue, free reign can be given to our moral imagination and ultimately to what we are here calling our underlying moral competence. While some cultures may deny that a woman accused of adultery has a right not to be stoned to death by her husband and other relatives, security against violent attacks is something people from all cultures generally recognize and see the point of. Violent threats to personal security, understood from the inside, are the same sort of immediate negative human experience for everyone, and protection against such threats would seem to be a naturally recognizable moral good. When you are minding your own business, stones aimed at your head are stones aimed at your head.

What are we to say, then, when respectful dialogue leads at least some women to claim that social protection against being stoned to death is not a moral good? Meyers (1997) builds on the work of earlier feminists in trying to understand how to make sense of the angry responses of some women towards actions or situations towards which other women remain acquiescent. Meyers notes that moral perception is a complex, many-layered process, prey to many distorting factors, including anger and bitterness. But anger and bitterness may also sharpen our moral perception. In the kinds of cases at issue, Meyers suggests that we should ask how a man would feel, were he to be treated in a similar fashion to the angry woman. If he too would feel angry, there is reason to believe that women's anger at such treatment is a clear moral perception that something is wrong with their social group's orthodox moral categories. Whether or not anger is justified in such cases is best ascertained by talking through the experiences of the angry person, sensitive to the facts that accusations of socially inappropriate anger are a significant means of silencing members of subordinate groups and that unquestionably acquiescing to one's subordination may often be the wisest course of social action.

Given their unorthodox nature, we should not initially expect heterodox moral perceptions to be widespread. So where do they come from, why do they persist, and why do they manifest themselves in anger or bitterness? And why do they win out, when they do? On our view, anger and bitterness are typical primate responses to situations where the individual involved is not being treated fairly relative to other individuals in the group. As we have argued in earlier chapters, some things are just not fair, and humans, together with other primates, seem to be biologically primed to recognize and care deeply about this kind of fact about their social worlds.

Heterodox moral perceptions are one important way in which a biologically based moral competence may affect processes of wide reflective equilibrium in humans. With regard to the wife stoning example, we might suppose that there is an impartial aspect to our underlying moral competence that enables us to correct the performance-based moral error involved in this form of apparent moral partiality. We call the error performance based because it arises from a process of social evolution that creates the social meanings that make wife stoning seem, at the level of actual moral performance or practice, like a morally reasonable thing to do.

In the remainder of this chapter, we examine several other more systemic forms of performance-based error that are likely to be more theoretically interesting and important than more specific instances of performance-based errors. As in the preceding example, which has of course its own systemic aspects, the forms of error we are interested in involve processes of wide reflective equilibrium and questions of when partiality is morally reasonable and when it is not. Because partial relationships are a central feature of human life, as well as the lives of other primates, we might suppose that they will be of central importance for a theory of human moral competence based in evolutionary biology.

Moral competence and moral performance

Systemic performance-based errors are generally understood to be important for understanding human psychological capacities. Our capacity for probabilistic reasoning, for example, seems to be structured in part by rules of performance that can sometimes lead to error: following these rules sometimes leads us to get things wrong in ways that our underlying competence can correct for (Kahneman, 2013). A good example is the fundamental attribution error (our discussion here is drawn from Andrews (2002)). People seem to be much more willing than they should be to attribute to a person's character actions that may have more to do with the situation in which the person finds himself or herself. For example, suppose we are told that an essay we are reading in defence of Fidel Castro is the result of a professor's assigning a student to write it. Once we are told this, we should be much less prone to believe that the student himself or herself has a pro-Castro attitude. Although people are indeed somewhat less prone to attribute a pro-Castro attitude to the student assigned the essay by a professor, they are not as prone to do so as they are when their getting the student's attitude right is crucial to their own success at some assigned task of their own. This suggests the fundamental attribution error is indeed an error, a performance-based error that can be corrected by our underlying competence should the circumstances warrant it. The error persists because it is usually benign: people typically act in accord with their beliefs and values, and so we are most likely to get things right if we assume a high degree of authenticity in judging people by their actions.

Systemic performance-based errors may help us understand how our minds have evolved. To begin to see how this might be true for our moral competence, let us turn to the question of how primates less sophisticated than we are approach some of their own moral dilemmas involving morally interesting forms of partiality. One important dilemma for almost all primates is reconciliation with one another after aggressive encounters. An assumption that runs throughout the work of Frans de Waal is that there will be regular clashes between individual and group interests in any species that is both social and intelligent. The interesting question for comparative psychology is how the members of the species in question navigate their way through or around such clashes in ways that might minimize the damage to either sort of interest. When and how do the clashes arise, and more importantly, how are they successfully resolved with minimal damage to all those involved? Does their resolution depend only on the individuals directly involved, or can it involve others, up to and including the entire group?

Among male chimpanzees, there is constant jockeying for power (de Waal 1989, 35–87 and 89–141). The alpha male in a group is typically supported by allies, and alliances are always open to shifts in loyalties. Male chimps fight more frequently than female chimps and, consequently, they also reconcile much more frequently than females. There is a corresponding difference in intra-sex cooperation: males tend to cooperate much more on a tit-for-tat basis, whereas females base their cooperation on bonds of kinship and close social bonds, bonds expressed in part by affiliative behaviour such as sitting together and grooming one another (49–50). De Waal puts these differences pointedly as follows:

> Male coalitions are instruments to achieve and maintain high status. There is little room for sympathy or antipathy in such opportunistic strategy. . . . Adult females, in contrast, live in a horizontal world of social connections. Their coalitions are committed to particular individuals, whose security is their goal.
>
> (51)

Rhesus females, on the other hand, live in a much more vertically oriented world. Rhesus society is made up of strictly ranked matrilineal lines, with all the females in higher lines outranking all the females in lower lines. But again, rhesus males reconcile, on the whole, much more frequently than rhesus females. This difference almost disappears, however, if one controls for kin and class relationships. Within their matrilineal groups and within matrilineal groups close to one another in the overall group hierarchy, rhesus females reconcile with one another almost as often as rhesus males do. To again quote de Waal,

> In a well-established social network such as a large breeding group, females concentrate on spheres of interest; they make up principally with their relatives and members of their own social class. So both sexes seem to do what serves them best in the natural situation, in which males wander from group to group and females stay in stable societies for their entire life (sic).
>
> (125)

Yet both chimps and rhesus monkeys seem to have some understanding of impartiality. When breaking up disputes over food, alpha chimps will typically prefer the underdog, even when the aggressor is an ally; moreover, alpha males who fail to prefer the underdog risk losing the support of the older females in the group. Also, male coalitions are changeable: your foe today may be your friend tomorrow. For male chimps, it is better not to be too partial. Though rhesus monkeys are much more rigidly hierarchical than chimps and seem never to prefer the underdog in their disputes but always their own kin or social allies, it seems that they too are capable of impartial behaviour:

> Several monkeys were trained to pull chains for food. After they had learned this response, another monkey was placed in an adjacent cage; pulling the chain now also caused the neighbor to receive an electric shock. Rather than pulling and obtaining the food reward, most monkeys stopped doing so in sight of their mate's suffering. Some of them went so far as to starve themselves for five days. The investigators noted that this sacrifice was more likely in individuals who had themselves once been in the other monkey's unfortunate position.
>
> (104)

This behaviour is particularly striking for rhesus monkeys that fight frequently and fiercely. While chimps' aggressive encounters are often more limited, physical violence from dominants towards subordinates is a frequent occurrence in rhesus behaviour, with dominant females doing a good deal of the biting.

Partiality and impartiality both seem important in the development of morality. How might they be related, from a biological point of view? Given the importance in primate evolution of female affiliation, and probably before it, maternal sensitivity, these particular forms of partiality might be supposed to be among the earliest forms of moral instincts in primates. Impartiality might then be supposed to be a secondary overlay on these earlier and more primitive forms of moral instinct: more primitive not in the sense that they are somehow less important, but in the sense that they might be the first and deepest part of human moral competence. If partiality is the earliest and deepest aspect of our moral competence, we might wonder whether impartiality evolved, at least in part, as a control mechanism to regulate partiality, or whether such regulation is merely something for which impartiality proved to be useful, once it had evolved for other reasons, such as coalition building among males. In either case, the result might appear to be disappointing for defenders of the foundational importance of moral partiality.

Little (1995) discusses impartial and partial viewpoints as sources of moral knowledge, drawing on some of the same feminist literature as Meyers. Like anger, other emotions (like caring) are often thought to distort our moral thinking; hence, to avoid such distortion, we should strive to take a more detached viewpoint in situations of moral conflict. Little's response to this line of thinking is that emotional capacities like loving care and affection can sometimes enable us to see things that would not be apparent to a more impartial observer (118). Little gives some nice examples in support of this point but raises the worry that one might respond to these examples on behalf of ethics as it has been more traditionally understood by arguing that all they show is that the moral perceptions of the caring person must sometimes serve as inputs to the process of moral reasoning undertaken by a more impartial observer. Little considers this point and thinks it mistaken:

> Affect serves as a helpmate to reason as he struggles with his imperfections. . . . [A]ffect is acknowledged as valuable for the aid she gives, but the value is only instrumental. . . . From a feminist perspective, of course, such a view has a depressing familiarity: once again it is what is associated with man that defines the ideal. I want to argue . . . that this move is not just depressing; it is wrong.
>
> (125)

Little argues that moral emotions may plausibly be claimed to have a kind of cognitive component that would make them a necessary part of moral knowledge itself. On her view, this may make moral realism true. While her argument remains vague about how and why this might be so, EMR begins to fill in the sort of empirical background that could tell us how and why we humans wind up with affective moral states the cognitive content of which matches up in direct kinds of ways with certain regularly recurring structural features of our social environments.

But here we want to focus on the view that Little rejects, the helpmate line of thought according to which moral emotions may simply help to supply appropriate inputs to our underlying human capacity for impartial moral reasoning. According to this view, affective attention focused on those we care for would be an epistemological parallel to the fundamental attribution error: although it is often useful as a heuristic moral device, it can also be an important source of moral error, measured against the corrective judgments of an underlying moral competence that is completely impartial in its underlying structure. Mill (2019, 220) seems to have some such view in mind when he argues that most of us do best most of the time when we intuitively prefer the interests of others who are close to us than the greatest good for the greatest number. For a utilitarian like Mill, this kind of partiality can make for good rules of moral performance, even if our moral competence is best modelled by the idea of the ideal moral observer who would count all interests equally.

From the point of EMR, we think that affective attention directed at particular others is unlikely to be simply a morally important kind of heuristic device and thus simply an aspect of moral performance. While we think that it can be a source of performance-based error, we also think that it is more likely to be an aspect of moral competence, balancing, and complementing impartiality. This is primarily for two reasons, both tied up with maternal and parental care more generally. First, caring for the interests of our children requires us to recognize and be moved by interests other than our own. On the other hand, our children are located particularly close to us as others who need ongoing and particular forms of help in our social environment. Their interests might be expected to loom particularly large on our moral horizon. But biologically, these aspects of the moral would have arisen in the environments of organisms incapable of distinguishing between self and others, never mind between the interests of self and others. Opportunities to help others would have been present in the environment, but the most important class of such opportunities would have been focused on more proximate others.

Second, to jump a long distance ahead in evolutionary development, humans, females and males alike, likely evolved as alloparents (Hrdy 2009). This suggests that impartiality does not start with but is already there to be exploited in male coalition patterns, at least in humans.

While feminist ethics has been important in highlighting the importance of considerations of care-based partiality in contemporary moral and political theory, we think that caring and impartiality are not fundamentally opposed to one another but are instead closely bound up together at the very foundation of the human capacity for morality. The central importance of impartiality in traditional moral and political theories reflects, on our view, a deep aspect of our human moral competence. The importance of impartiality does not simply emerge along with human reason and the realization that social groups can be run more evenly with at least a veneer of justificatory reasons that are largely impartial in their appeal to all the different members of any given social group. We think care-based partiality to the needs of others in particular webs of relationship reflects a second and similarly deep aspect of our human moral competence. It does not simply arise, for example, as a socially based structural element in the thinking of the members of subservient social groups that are charged with directly caring for the physical and emotional needs of others, a possibility explored in a feminist context by Harding (1987).

One way to model these twin aspects of human moral competence is on a continuum. Fundamentally, the interests of others matter to us. But these interests may be the interests of others who are closer to or more distant from us. As the moral situation demands, we may slide more or less in one or the other of these two directions: that of greater partiality or that of greater impartiality. This may be the same general sort of thing that is going on with rhesus females and chimp males at a more instinctual level: because of the two different kinds of moral contexts they find themselves in, rhesus females, on the one hand, instinctually slide more in the direction of partiality in reconciling with other females, while dominant chimpanzee males, on the other hand, instinctually slide more in the direction of impartiality when breaking up altercations that one of their coalition allies is involved in.

As we saw in the last chapter, the human capacity for empathy is much more complex and likely to be much more open-ended than empathy in other primates. John Brown's own empathy for the plight of slaves arose in part from his own partial relationship to another young boy, but it readily expanded to encompass other young slaves and then slaves more generally. While the empathy of other whites was clearly limited by racial prejudice, a certain amount of this prejudice was overcome by the course of the war and the acts of the freed slaves who fought bravely for the North. Not all prejudice, or even most of it, as the next century of racial intolerance towards blacks was to make painfully clear. Partiality to those we perceive as being closest to us can sometimes be a deeply rooted and socially disruptive source of performance-based moral errors. But sometimes not: sometimes it is right to care for those close to one in ways that one would not be expected to care for others who are more distant. Impartiality and partiality both seem to be deep structural aspects of human moral competence.

But it also seems true that partiality can collide with impartiality, and that impartiality can act as an important regulatory mechanism against too much or the wrong kind of partiality. At the level of human reasoning about morality, the important moral questions are whether and when it should. Such questions are likely to be vexed because both partiality and impartiality seem to be equally deep aspects of our human moral competence. How might our underlying moral competence affect our moral reasoning, when the two kinds of moral reasons clash?

One of the main problems with partiality as a moral motive or reason is that the collective pursuit of partial moral aims can be self-defeating. This is just as likely to be true at the biological level as at the level of human social institutions. Regarding this problem at the social level, Parfit (1984, Section 36) points out that "each-we" moral dilemmas can arise whenever human moral codes assign moral agents individually different moral aims. One such aim might be to take some sort of special care of your own relatives. At a general level, this is a universal moral obligation, but for each individual the obligation is aimed at a different set of particular others. Each-we dilemmas arise in cases where if each of us does what is partially required of us, we together all do worse in terms of the same set of obligations. If each of us reasons in a partial way and tries to act directly on our obligations to do what is best for our own relatives, we may each do worse in fulfilling this obligation than if we had reasoned more impartially. Each-we dilemmas can only be resolved if those involved reason in a more impartial way about the partial values at stake. In this way, impartiality can function as an important mechanism for furthering or protecting moral values arising from relationships of partiality. The open-ended aspect of our empathy can enable us to repair or avoid an important sort of performance-based error that stems from focusing too much on the interests of particular others in particular relationships to us. Empathy can thus drive impartiality at the level of moral performance, helping us not to focus on the interests of those too similar or too close to us: but not always.

Almond (2005) considers a good example of the sort of performance-based error at issue here in a situation where one desires the last available life jacket for one's own child. If there is only one life jacket and more than one parent and child, parents may fight over the life jacket and no child may be saved. Or if we have enough life jackets, but I am closer to your child while you are closer to mine, we may save neither child by both trying to get life jackets for our own children. Problems with research ethics boards notwithstanding, it would be interesting to see how parents generally react in such rare emergency situations. One supposes that in such situations, most parents would be prone to a particular kind of performance-based error, that of preferring the interests of their own children to the interests of the children of others. Like the fundamental attribution error, this mode of reasoning would be, as a matter of fact, an error: an error that our competence enables us to identify as such when we think about such situations more reflectively. Each of our children would do better, were we not to prefer the interests of our own particular children. But the error here is an error for reasons of both impartiality and partiality. The harm of my child doing worse in this situation is closely connected to the harm of all the children doing worse. Saving fewer children is a grievous moral error, one we might grieve over both separately and together.

Unlike the fundamental attribution error, preferring the interests of our own children is not a simple heuristic device that enables us to satisfy our impartial aims better than we otherwise might in the context of general kinds of situations we often find ourselves in. The error is a tragic moral error because when we commit this kind of performance-based error, we lose not just more lives but the lives of particular individuals whose well-being deeply matters to us. To the extent that we accept such tragic outcomes as the result of our doing the best we could in a difficult situation, this might be because cases of the kind in question arise unexpectedly and are relatively rare.

In cases where this kind of error is recurrent and predictable, we might expect our underlying moral competence to lead to forms of moral reasoning that would block it. This is what makes the error an error rather than simply an inevitable and tragic aspect of the moral life. Consider an example of an each-we dilemma from the anthropology literature (Stingl 1996). The Karimojong are a Nilo-Hamitic tribe living north of Lake Victoria in eastern Africa. In their traditional way of life, the people of this tribe lived in permanent settlements during a lengthy wet season of heavy rain. During the dry season, however, when there was not enough water in settled areas for both people and cattle, the herdsmen of the tribe would have to leave the settled areas with their cattle in search of water. This could result in the following kind of situation:

> herdsmen who meet at one watering place will come from many different settlements, and no one will expect to meet the same people each year. . . . The Karimojong realize this and say, "The sun mixes us up." They are most mixed up at the height of the drought, when a number of herds and their herdsmen combine to use the same water and grazing and to keep others out of it. If a conflict of this kind occurs, loyalties are clear. The "insiders" in this temporary group must stand together against the outsiders, whatever ties of kinship or neighborhood may bind them to the outsiders at other times.
>
> (Mair 1970, 25–26)

In this case, impartiality prevents partial moral reasons from becoming unreasonable.

Such cases are not odd. Consider another case, where the approach to evolutionary ethics we are developing here predicts that although we will eventually do what we ought to, getting to this point will take some time, because it will require us to correct the powerfully attractive performance-based error behind Almond's life jacket example. Current medical practices will not typically allow organ donations when family members, after the death of a loved one, refuse to agree with the documented wish of their loved one to offer his or her organs to others. Here our current considered judgments respect the partiality such family members often appear to be acting on, as they try to do everything they possibly can to protect the bodily integrity of a loved one in a vulnerable position, up to and including brain death. But such partiality cuts both ways, depending on whether a loved one is a potential donor or a potential recipient. The more regularized our current shortages of donor organs become, the more likely it is that our current reflective equilibrium will shift in the direction of the Karimojong. (For a widely noted proposal in this direction, see Spital (2003).)

In an evolutionary context, it is useful to compare these sorts of cases to male chimps and their allies and to contrast them with the relative inability or unwillingness of female rhesus monkeys to reconcile across established social strata. If each alpha male favours his allies in disputes over food, the alliance does worse, in the long run, than if these allies had not been favoured. Alpha male favouritism leads the older females to step into the fray, threatening the position of the alpha male and hence his alliance with his current favourites. In certain regularly occurring kinds of situations, partiality fails as a moral good on its own terms, and in these situations, a more impartial approach to reasons of partiality is morally better than an approach that is entirely partial. The relative inability, or unwillingness, of rhesus females to reconcile outside their social cohort is in all likelihood not self-defeating in their normal ecological context, but that context could change in ways that might make their current level of partiality self-defeating. Their moral instincts might enable them to detect and respond to the fact that this is so or they may not, depending on just how sophisticated rhesus moral instincts turn out to be. And depending on the extent of the ecological change, rhesus monkeys as a species might be in more or less trouble. Something similar may be true for us, a point we take up in the next section. In the face of threats like global warming, humans may need to transcend national boundaries in cosmopolitan ways that we currently seem hesitant to pursue.

We end this section with the important note that the kinds of performance-based errors we are talking about in this chapter can arise in the direction of either of the two aspects of our moral competence we are discussing here. Just as too much or the wrong kind of partiality can lead us into correctable moral errors, so can too much or the wrong kind of impartiality lead us into correctable moral errors. In both cases, there may be systemic forms of performance-based error well worth investigating within the context of the general approach to ethics we are developing in this book. One could thus read much of the critical feminist literature on traditional ethics and social and political philosophy as an exploration of systemic forms of performance-based moral error based on failures in our moral reasoning at the level of what we might be otherwise prone to accept as morally right on the basis of impartial moral reasons. We have not explored this rich path of argument in this chapter because from the perspective of an approach to ethics based on evolutionary considerations, impartial moral reasons might initially seem to be a much more likely source of moral error from the point of view of protecting and maintaining the sorts of relationships that ought to matter most deeply to us. But further development of our general approach to human moral competence would require more detailed attention to both sources of performance-based error.

Human impartiality and tribalism

As a psychological capacity, human moral competence is obviously much more sophisticated and much more internally adaptable to new moral situations than the moral instincts of rhesus monkeys. Human social networks, unlike those of other primates, appear to be able to expand without limit, and both partiality and impartiality appear to play important roles in such processes. As human societies expand, they become segmented in ways that build upon the simple social hierarchies of other primates, such as the rhesus females that are situated towards one another in both familial and class relationships. Segments of a larger group can themselves be segmented, and individual obligations can travel up and down the segments in such a way that one might be simultaneously obliged to attack or to defend a particular individual or segment depending on the social context in which those involved must interact. To take another dramatic case from the anthropological literature, in traditional Nuer society an individual might be socially expected to attack someone from another segment of the tribe, perhaps because of an earlier cattle theft (Evans-Pritchard 1940, Chapter 4; Stingl 1996). But at the same time, he or she might also be obligated to defend that other individual, were both to be attacked by individuals from outside the larger segment that joins both their more immediate segments.

This sort of case helps to explain, if not to resolve, an important difference between Miller (2005) and Bader (2005) regarding the strength of our moral obligations to fellow nationals. Why, unless the situation is dire, might nations justifiably choose to benefit their own citizens rather than the citizens of other nations who are much worse off? Partial relationships are typically seen by the individuals involved in them as being, at their core, intrinsically valuable, and the uncontracted obligations that are attached to such relationships are typically seen as an important feature of this intrinsic value. According to Miller, this much is certainly true of familial relationships; the question is whether the same point can be extended to fellow nationals, and hence whether nationalism is also a form of reasonable partiality. Miller thinks that it is. While nations are hardly families, we might point out in defence of Miller that nations are currently the largest social segment linking modern individuals to one another in a generally dependable sort of way. As social segmentation expands, having now grown to a national level, it brings with it significant changes in social identity, trust, and motivation. But just as we can always expect to find tensions among individuals and their groups, we can also expect to find tensions among segments within and across levels. Such relationships will always be to some degree conflicted, but what is important is that when the larger segments demand trust and motivation to protect segmental values against internal or external threats, enough trust and motivation exist for the requisite level of segmental solidarity.

Against Miller, Bader claims that such solidarity is precisely what we no longer have at the level of the nation state. Globalization is pulling the nation state apart from the outside while internal economic and ethnic differences are pushing it apart from the inside. But here it is important to consider that as social networks grow in size, we might expect both normal and revolutionary periods in social segmentation. During revolutionary periods, as larger social segments are forming, we might expect, as a particularly important kind of performance-based error among humans, resurgences of internal segmental conflicts, resurgences of what might be called the worst aspects of tribalism. This will cause the former largest segmental relationship, in this case nationalism, to appear, as it does to Bader, as too weak a moral and social bond to move us forward, either in the direction of acknowledging and shouldering our emerging global moral obligations or in the direction of resolving and discharging our contested moral obligations to our fellow citizens.

From a biological point of view, earlier segments will typically remain important to individuals, even after these segments have been encompassed by newer and larger segments. Nevertheless, the larger the segment, the more it might be expected to be pivotal in times of revolutionary social growth.

Bader and Miller each are thus half right and half wrong. In an unstable, globalizing world, the smaller segments that make up nations will reassert their importance because of their longer term stability in terms of identity and trust. On the other hand, national segments will be pivotal in the process of globalization, because they will be what the new, more encompassing segments are most directly built upon. In both cases, however, hanging on too tightly to the hitherto reasonable partiality of more limited segmental relationships can easily become a performance-based error. When segmental growth is successful, an important part of its success must be found in our underlying ability to correct precisely this sort of error. Correcting this kind of error does not eradicate earlier segmental obligations, but it does redraw their limits in a way that allows for new levels of social identity, trust, and motivation to emerge. While earlier forms of partiality remain reasonable, the limits of reasonable partiality significantly change, contracting at older levels and expanding at new and higher levels of social organization.

What this means for the debate between Miller and Bader is that during a revolutionary period of social growth, there will be no uncontested considered moral judgments to appeal to regarding the reasonable limits of national partiality. Without an underlying moral competence to push us past such moral conflicts, we would have remained mired in our tribal past. While the debate between Miller and Bader is not immediately resolvable by appeal to any of the fixed points of our current wide reflective equilibria, East, West, North, or South, we might thus continue to hope, on the basis of our moral competence and our past episodes of successful social growth, that the limits of our current forms of partiality will be redrawn in a way that allows for more global forms of social segmentation to emerge.

What this means for a theory of our moral competence is that while systemic forms of performance-based error can tell us important things about the biologically based structure of this psychological capacity, the fact that we are ultimately able to correct such errors can tell us even more. In the case of human moral competence, impartiality would seem to be an important check on more partial approaches to the needs and interests of others. At least, this seems to be true for each of the three examples we have considered in this chapter: heterodox moral perceptions regarding wife stoning, each-we dilemmas involving family members, and the limits of older categories of reasonable partiality in a globalizing world.

Group loyalty

Let us return, in conclusion, to the example of moral partiality with which the book began: Canadian soldiers lying buried in Flanders' fields. Group loyalty, along with trust, seems likely to be a natural moral value for a view like EMR. It remains an open question, however, how this natural moral value might be tied to claims about whether we ought to join a particular war effort to continue the fight of those who fell before us or whether we ought to resist this further call to arms. Complex moral conclusions like this depend on complex moral arguments that take place within processes of wide reflective equilibrium that include a host of considered moral judgments as well as empirical judgments about the surrounding facts that pertain to the situation in question. We think that natural moral facts and our human moral competence generally affect the moral conclusions we reach but that their causal effects are mediated by processes of wide reflective equilibrium. Exactly where, how, and when this happens are difficult questions to answer. We have made a start at answering such questions in this chapter and the last but only a start. Our argument in this book is that EMR is a promising theory of morality. Exploring this promise more fully will require more detailed efforts to develop the theory.

Note

1 An earlier version of this chapter appeared as "Reasonable Partiality from a Biological Point of View," *Ethical Theory and Moral Practice* 8 (2005): 11–24. We would like to thank Springer Nature for permission to reprint this material here.

Bibliography

Almond, Brenda. 2005. "Reasonable Partiality in Professional Relationships." *Ethical Theory and Moral Practice* 8 (1–2):155–168.

Andrews, Paul. 2002. "The Psychology of Social Chess and the Evolution of Attribution Mechanisms: Explaining the Fundamental Attribution Error." *Evolution and Human Behaviour* 22:11–29.

Bader, Veit. 2005. "Reasonable Impartiality and Priority for Compatriots: A Criticism of Liberal Nationalism's Main Flaws." *Ethical Theory and Moral Practice* 8 (1–2):83–103.

Cohen, Joshua. 1997. "The Arc of the Moral Universe." *Philosophy and Public Affairs* 26 (2):91–134.

Daniels, Norman. 1979. "Wide Reflective Equilibrium and Theory Acceptance in Ethics." *The Journal of Philosophy* 76 (5):256–282.

Daniels, Norman. 1980. "Reflective Equilibrium and Archimedean Points." *Canadian Journal of Philosophy* 10 (1):83–103.

de Waal, Frans B.M. 1989. *Peacemaking among Primates*. Cambridge: Harvard University Press.

Drydyk, Jay. 1997. "Globalization and Human Rights." In *Global Justice and Global Democracy*, edited by Jay Drydyk and Peter Penz, 159–183. Winnipeg and Halifax: Society for Socialist Studies and Fernwood.

Evans-Pritchard, E.E. 1940. *The Nuer: A Description of the Modes of Livelihood and Political Institutions of a Nilotic People*. Oxford: Oxford University Press.

Harding, Sandra. 1987. "The Curious Coincidence of Feminine and African Moralities: Challenges for Feminist Theory." In *Women and Moral Theory*, edited by Eva Feder Kittay and Diana Tietjens Meyers, 296–315. Savage, MD: Rowman & Littlefield.

Hrdy, Sarah Blaffer. 2009. *Mothers and Others: The Evolutionary Origins of Mutual Understanding*. Cambridge, MA: Harvard University Press.

Kahneman, Daniel. 2013. *Thinking, Fast and Slow*. New York: Farrar, Straus and Giroux.

Little, Margaret Olivia. 1995. "Seeing and Caring: The Role of Affect in Feminist Moral Epistemology." *Hypatia* 10 (3):117–137.

Mair, Lucy. 1970. *Primitive Government*. Harmondsworth: Penguin.

Meyers, Diana Tietjens. 1997. "Emotion and Heterodox Moral Perception: An Essay in Moral Social Psychology." In *Feminists Rethink the Self*, edited by Diana Tietjens Meyers, 197–218. Boulder: Westview.

Mill, John Stuart. 2019. *The Collected Works of John Stuart Mill: Essays on Ethics, Religion, and Society*. Vol. 10. Toronto and London: University of Toronto Press and Routledge and Kegan Paul.

Miller, David. 2005. "Reasonable Partiality towards Compatriots." *Ethical Theory and Moral Practice* 8 (1–2):63–81.

Parfit, Derek. 1984. *Reasons and Persons*. Oxford: Clarendon Press.

Rawls, John. 1971. *A Theory of Justice*. Cambridge, MA: Harvard University Press.

Spital, Aaron. 2003. "Conscription of Cadaveric Organs for Transplantation: Neglected Again." *Kennedy Institute of Ethics Journal* 13 (2):169–174.

Stingl, Michael. 1996. "Evolutionary Ethics and Moral Theory." *The Journal of Value Inquiry* 30:531–545.

8 Moving from is to ought

The naturalistic fallacy

In proposing an evolutionary origin for morality, the general line of argument in this book may seem to have moved from is to ought, from the way things are in world to the way things ought to be. Doing this in the wrong sort of way, or perhaps in any way at all, is sometimes called the naturalistic fallacy. That something is natural gives us no immediate reason to think that it is good. We cannot validly reason from claims about the way the world naturally is to claims about the way the world morally ought to be.

According to EMR, morality first arises with simple moral values, morally good kinds of things that arise in certain kinds of cooperative environments. What makes something a moral good is the kind of thing that it is, similar to things like ducks: what makes a duck *a duck* is the kind of thing that it is. Moral instincts arise as evolved (and evolving) mechanisms for internalizing external moral values in motivationally efficacious ways. Moral competence, in humans, arises as a more highly developed innate psychological mechanism for producing moral judgments about the world that same competence enables us to recognize as moral reasons for our subsequent actions.

Moral reasons are grounded in moral judgments, which judgments, minus any background noise from such sources as religion, self-interest, and hegemonic class interests, are ultimately grounded in natural moral values, along with the internal structure of our moral competence and our participation in psychosocial processes of moral wide reflective equilibria. Not all moral reasons so produced will be equally reasonable, but to the extent that moral reasons are attached to moral judgments produced through a process of wide reflective equilibrium, they will have morally normative force, a force grounded in our moral competence and the natural moral values it has developed in response to, amplified by the force of reason that seeks to make all of our judgments consistent with one another and the force of our shared social commitments to a joint process of deciding how to best structure our relationships with one another.

Because EMR is an empirical theory, the general form of our argument for EMR is abductive. We start with the apparently real force of moral normativity and ask what would have to be true in the biological world to explain the existence of this form of normativity. At no point in our argument do we offer a deductive argument from statements about facts to statements about values. Instead, at each step of the argument, we maintain that at their most foundational levels, moral values and moral normativity are best understood as being built into the biological facts that our overall argument for EMR rests upon. Moral values are things in the world that allow for the development of moral circuits linking these values to positive or negative responses from creatures for whom these values matter, responses that vary depending on whether the natural moral values causally tied to them are positive or negative, like helping or cheating. These circuits allow for the development of instincts, which further respond to the world in attempting to alter it in positive directions determined by those same instincts.

In humans, moral competence emerges as a cognitively more sophisticated normative capacity, a capacity that responds to moral aspects of the world in the course of trying to make the world better fit the ways in which the capacity determines it ought to be structured. As we began to develop the idea of the human capacity for morality in the preceding chapter, we identified human moral competence as a psychological mechanism for both detecting moral features of the world and effecting positive moral changes in the world. We think this mechanism is linked to our capacities for thought and language, and thus also linked to moral argument and agreement. Drawing on our underlying moral competence, our capacity to reason, and our ongoing social commitments to long-term projects and to each other, moral arguments seek to determine how best to proceed in making the world more like it morally ought to be.

EMR thus distinguishes between morality more generally and human morality more specifically. Biologically, morality begins with natural moral values and the organic circuits that arise in response to these values in organisms for whom these values matter. The specific form of morality that has evolved in humans further depends on human moral norms that are articulated and agreed upon in the context of human thought, language, and argument. In claiming that morality in general is normative, EMR is claiming that moral values and the circuits they become part of have a natural to-be-doneness written into them. In making this claim, EMR directly opposes, on empirical grounds, Mackie's (1977, 38–42) metaphysical argument that moral properties, if they really existed, would be a bizarrely anomalous feature of the real world of material objects, insofar as they would have to be supposed to be at once both physical parts of the world and the sorts of things that would somehow tell particular kinds of evolved organisms that happen to bump into them what to do. To use Mackie's own phrase, this would indeed be queer.

To respond to Mackie's argument, we need to distinguish between morality as a broader biological phenomenon and human morality in particular. Human moral norms tell humans what they ought to do, but natural moral values and the organic circuits they are part of are not moral norms in this sense of the term, and in this sense of normativity they tell no one what he, she, or it ought to do.

Human moral imperatives depend on human moral judgments that represent expressly articulated sets of moral norms. To be clearly understood and effective, moral norms need an ongoing human context of language, argument, and agreement. In supposing there are biological moral imperatives in the more general sense of moral normativity, EMR is not referencing moral imperatives in this specifically human sense, although it is supposing that natural moral values and the circuits they are part of can be loosely understood in an imperative voice. To return to our discussion at the end of Chapter 2, organisms actively look for food. In eating something that satisfies the need for food, an organic circuit can become established linking particular sorts of stimuli in an organism's environment to particular sorts of responses in the organism. In effect, when the organism encounters the right kind of stimulus, an "Eat this!" response is already there to be triggered. Organisms are wired to be looking for the kinds of things they can eat. Although there is always a chance element to any such encounters, organisms are built to be actively looking for things that complete the biological circuits, or imperatives, they are operating under. So when they bump into something that it is indeed good for them to eat, this is not entirely a chance encounter or a stimulus looking to trigger an interesting response. Organisms are themselves already looking for the right kind of stimuli, whether this involves nutrition or morality. The underlying idea is that organisms evolve as active, striving agents within their environments, not merely as passive transfer mechanisms that happen to link certain stimuli to certain responses.

An example of this general phenomenon can be observed in the online flood of cat-and-cucumber videos. Cats are actively interested in both prey and predators in their environment as distinct kinds of things. If you place a cucumber behind a cat when it is otherwise occupied, say, at its food dish, when the cat turns around to find the cucumber behind it, it will leap high into the air and well out of harm's way. While it is unlikely that the cat is frightened by the cucumber itself, it is certainly frightened by the kind of thing the cucumber might be.

In a more static world of billiard-ball causes and effects, where stimuli happen to sometimes cause responses and certain kinds of responses happen to get associated over time with certain kinds of stimuli, any objective properties of to-be-doneness would indeed be highly dubious sorts of things. All that would exist in this vicinity of the material world would be associative learning patterns and related behaviours. But in a dynamical biological world where organisms are actively seeking to do certain kinds of things, like eat or avoid being eaten, certain kinds of stimuli and the organic circuits that they become part of will have written into them a natural aspect of "do this kind of thing with that kind of thing." In this sort of dynamical biological world, organisms are out to do certain kinds of things, and organic circuits are waiting to be formed that will tell them, once formed, what they ought to do when, in what we might call a biologically imperative sense of ought. The nature of these oughts will vary depending on the nature of the objects that trigger them, for example, whether these objects are nutritional goods or moral goods. A moral analogue to "Eat this!" would be something like "Provide help!" If we're thinking of porpoises, the first imperative might be triggered by a fish and the second by a floundering, air-breathing figure near the surface of the water.

Yet even in a biological world that is assumed to be more static than dynamic, natural moral values would not simply be physical or chemical arrangements of particular atoms or molecules. They are things that come into existence as the particular recurring arrangements of objects that they are in the environments of particular kinds of evolving organisms. Like predators, they are good things to pay attention to even prior to the formation of an appropriate circuit, such as "Run away!" or "Provide help!" There is a biological "to-be-doneness" aspect built into them from the very start of their existence, prior to the development of appropriate responses to them in organisms for whom they matter, however we may suppose these responses to develop. The to-be-doneness of such things is part of what it is to be the natural kind of thing that they are, if EMR is right.

From biological circuits to moral norms

Having said all this, we still need to address more directly the question of whether the abductive nature of our argument really does enable us to avoid what we are calling the naturalistic fallacy. Here is one way of casting our argument that might make it look as if we have not really avoided the fallacy of deriving an ought-conclusion from factual premises:

(1) *Fact*: Helping those who need it is a natural attractor for organisms in certain kinds of cooperative networks.
(2) *Fact*: Humans have a natural moral capacity for noticing and positively responding to things like helping those who need it.
(3) *Fact*: In thinking and arguing about stoning women to death for adultery, humans in modern liberal societies have come to the considered moral judgment that it is morally wrong.

Therefore,

(4) *Moral ought*: Humans in modern liberal societies ought to come to the aid of women who face being stoned for adultery in societies that allow this.

For EMR, (4) is not a claim that floats freely from the process of wide reflective equilibrium that produced it. Like (3), it is contained in this process, as a considered moral judgment in wide reflective equilibrium along with other considered moral judgments and along with beliefs involving various empirical facts about the world, including biological, anthropological, sociological, and psychological facts as well as we are able to understand them. The highest form of moral ought that arises for humans at the point we are on, on our own moral trajectory, is the kind of ought that stems from our considered moral judgments in wide reflective equilibrium. There is no breaking outside of this process to find a deeper and more fully justified sense of moral ought.

The general idea of moral wide reflective equilibrium is that it contains all of a given society's moral judgments and related non-moral beliefs at a given point in that society's social evolution. What we are really dealing with are ongoing sets of shifting equilibria as different societies develop, refine, and change for the morally better or worse their entire sets of moral judgments. Such judgments will range from the highly general to the highly particular, from general principles like "never fail to respect the autonomous choices of others when they are doing the same" to "you should not have lied last night when your spouse asked you where you had been." Included as well will be a wide variety of mid-level rules, such as "respect patient autonomy" from institutions and professions charged with delivering health care.

Rawls (1971, 17–22 and 577–587) also wants, sensibly, to include our judgments about the contractualist apparatus he thinks is part of the justification of basic principles of social justice, elements of which include the idea of the Original Position and the Veil of Ignorance. Because we are including all judgments with moral content, this makes sense. How we think about impartiality, the moral point of view more generally, and what counts as a successful moral argument are all things that are part of a moral equilibrium point for any given society at any particular point in time. These latter elements of wide reflective equilibrium will have both empirical and moral dimensions, such as how, exactly, most of us view impartiality or the moral point of view.

The other thing making such equilibria wide is the inclusion of all relevant empirical theories, theories from psychology, sociology, and anthropology, as well as relevant historical beliefs and other beliefs about how humans work as individuals and in societies, beliefs coming either from the humanities or from social thought more generally. If a belief or theory is relevant to morality, it is included within the set of beliefs that need to be made consistent and coherent with one another, many of them moral judgments with oughts embedded in them. We assume that if EMR becomes an established empirical theory, it will similarly contribute to processes of moral reflective equilibrium.

Because moral ought-claims are always embedded in a wide reflective equilibrium, if they appear as conclusions of arguments meant to establish their truth, those arguments can always be represented as closed propositional circuits of claims drawn from within the wide reflective equilibrium in question. If they are to be successful, moral arguments must have this structure. EMR takes the normative aspect of natural moral values to be part of the natural world, and it also supposes that beliefs about these values and their normative force will naturally make their way into human moral thinking and human reflective equilibria. Some of the beliefs in such equilibria will be empirical in form, some of them will have morally prescriptive content, and some of them may well have mixed content. This is an important part of what makes moral reflective equilibria wide: they contain not just a human group's considered moral judgments but also all its judgments about the world that are relevant to those judgments. In this regard, premises (1)–(3) in the aforementioned argument could easily be supposed to be part of our own current wide reflective equilibrium. But there is much more to the argument for (4) then is contained in (1)–(3). These scant premises need to be combined with a lot of other considered moral judgments for us to get a justifiable argument for thinking (4) to be true.

For example, to return to our Chapter 5 discussion of wide reflective equilibrium and in particular to Scanlon's example of the general duty of rescue we might suppose ourselves to owe to one another in our current state of wide reflective equilibrium in Western democracies, we should once again note that while we do allow for a certain amount of rational self-interest in our current moral thinking, we do also think that rational self-interest should be constrained in ways that matter to the good of others even if there is nothing in it for us. Based on other considered judgments regarding when and how we ought to help others, it might appear that we have a prima facie reason for intervening in the stoning case, even if it brings us no benefits and, indeed, even if it might conceivably result in an uptick in terrorist threats against us. If the risk to us is relatively low, or we might be subject to it in any case, we should step in to help someone who is in great distress. But there are still other moral considerations: if the country in which the stoning is taking place is a sovereign nation, by what justifiable moral rules do we intervene in its sovereignty, as individuals, coalitions of individuals, nation states, coalitions of nation states, or any or all of the above? Should "we" be intervening at all, and if so, in what manners or degrees? The point is that we are never going to get from a simple natural value like helping others to an interesting moral conclusion, not outside of a multifaceted argument taking place within a particular wide reflective equilibrium of considered moral judgments.

Although the abductive argument of EMR does build a biological kind of moral normativity into what we are calling natural moral values, there is no direct connection from this biologically based form of moral normativity to the "oughts" of moral judgments in wide reflective equilibrium. These moral oughts will have to be logically supported by that equilibrium itself, which will always contain a variety of ought-statements relevant to the truth of the one in question, as well as a variety of relevant judgments not containing oughts that we also have good reason to believe. EMR adds to the overall stock of judgments that are not considered moral judgments empirical claims about the evolution of natural moral values. While EMR builds normative elements into both the values and the moral capacities that respond to them, the considered moral judgments that come out at the other end of moral arguments are never solely supported by these values or capacities. EMR does not logically derive moral "oughts" from empirical "is-es." On the other hand, as our slavery argument suggests, natural moral values and capacities may play a role in the formation of reflective equilibria whether we are paying explicit attention to them or not, and from this same example we might also suppose this role could be enhanced by more empirically developed forms of attention to the underlying moral values in question. The more we know about these values, the more solid our moral reflective equilibria will be, or so we might at least optimistically suppose.

What EMR proposes at its core is an empirical research program. We think this research program may generate interesting empirical results regarding natural moral values, moral capacities, and moral trajectories more generally. We also think that if such results are generated, they will most likely have effects on our considered moral judgments in wide reflective equilibrium. But we will never be in a position to directly read off a considered moral judgment from a natural moral value. This is our point about aforementioned premises (1)–(4). This argument schema is simply not an accurate representation of how justified moral judgments come into the world. More generally, if we were to ask evolutionary biologists for a list of justifiable moral judgments, we would come up empty-handed.

Can biology take over ethics?

In the first chapter of *On Human Nature*, E.O. Wilson (2004) infamously suggests that it is precisely to biologists and neuroscientists we should turn when it comes to determining the deepest aspects of moral justification and truth:

> In order to search for a new morality based upon a more truthful definition of man, it is necessary to look inward, to dissect the machinery of the mind and to retrace its evolutionary history.
>
> (4)

> innate censors and motivators exist in the brain that deeply and unconsciously affect our ethical premises; from these roots, morality evolved as an instinct. If this perception is correct, science may soon be in a position to investigate the very origin and meaning of human values, from which all ethical pronouncements and much of political practice flow.
>
> (5)

Key to Wilson's argument here is the idea that the set of considered moral judgments we wind up accepting at a particular point in time will be determined by what seems right to us, and what seems right to us will be determined by our brains, most particularly by our emotions and our limbic system. As he says about the rival political theories of John Rawls and Robert Nozick,

> Like everyone else, philosophers measure their personal emotional responses to various alternatives as though consulting a hidden oracle. That oracle resides in the deep emotional centers of the brain, most probably within the limbic system, a complex array of neurons and hormone-secreting cells located just beneath the "thinking" portion of the cerebral cortex.
>
> (6)

Put this baldly, Wilson's argument immediately seems to fall prey to the naturalistic fallacy as we are describing it here: things turn out to be right because, as a matter of fact, we think so, and we can better see what to think by studying the brain systems that produce these thoughts.

One might try to defend this argument against the fallacy by claiming that all there really is to moral rightness is what we think to be morally right, so there is really no illicit move here from psychological and neurophysiological facts to some form of a moral ought-statement but only the move from one class of factual statements to further factual claims about what we think ought to be the case or what we would think ought to be the case, were we to get the underlying science of our thought processes right. So we never get to true moral oughts in the conclusions of our moral arguments but only to conclusions about what we think we ought to do. As a crucial part of knowing what to think about morality in a reliably scientific sort of way, we consult the source of all human moral thought, the human brain and limbic system.

We might object here that there seems to be more to such conclusions than this – Wilson is, after all, presenting us with what he calls a new morality, a new morality based upon inescapable scientific truths about humans and the ways in which our brains have evolved. As such, this new morality is being pressed against those of us who might doubt its dictates as being what we ought to think about what ought to be the case. But then a defender of Wilson might just say that this second level "ought" is still captured within what we think ought to be true about what ought to be true. In arguing about morality, we can never escape the realm of our own thought processes.

Against this point, a contractualist might offer the argument that these processes are to be held to the objective standard of what truly could be agreed to, were we to reason in a way that genuinely and fully considered everyone's basic needs and interests. In this case, moral truth is being determined not just by psychological or neurological facts about humans but by social facts about what we would in truth agree to under certain conditions meant to ensure that we respect each equally other as separate individuals. But then we would need to know, it would seem, where the moral importance of the contractualist standard itself comes from. The human limbic system?

According to EMR, what all these arguments are missing is that the fundamental moral conditions of human interaction are set by the natural moral values that are part of our existence as the kind of biological species that we are. There may be natural moral values on our trajectory that are not on the trajectories of other species, and indeed may never be, but these specifically human natural moral values are of a kind, or perhaps kinds, with the natural moral values that are a part of evolutionary development of morality more generally. Human moral oughts are thus grounded in something real, something that exists outside human evolution, human thought, human psychology, human sociology and anthropology, and human neurophysiology. There is more to what we morally ought to do than what we might on reflection think we ought to do.

This point returns us to our main question of this chapter: if natural moral values are an important part of telling us what we ought to do, how does EMR avoid moving from facts to values? EMR does suppose that moral values and a general form of moral normativity are built into the natural world from the very beginning of moral trajectories. And it does suppose that these values will influence the direction of moral wide reflective equilibria. But to influence such equilibria, the moral values involved will have to be articulated in language, argued about, and agreed upon in the course of rationally deciding which moral judgments or norms belong to the equilibrium and which ones do not. There is no direct logical pathway from natural moral values to considered moral judgments in wide reflective equilibrium. And yet such moral values may exist, in causally potent ways, just as the earth may yet move, quite apart from what any currently accepted dogmas might tell us we otherwise ought to believe.

Along with rejecting Wilson's idea that our limbic systems are the foundation of morality, we should also reject Dawkins' related idea that our human moral thinking is always going to be bound by what is good for the survival and reproduction of our human genes. Given the fact that we are, as a biological species, on a moral trajectory, what is morally good for us is ultimately determined by the natural moral values that exist for us at the point of our trajectory that we are currently on. Paying attention to our own interests may have some moral value, as a natural moral value, but it is one natural moral value among many others. One of these other values will involve paying close attention to the interests of our offspring, but again, this will be one natural moral value among others.

In thinking about how much relative weight to give our own interests, or the interests of our children, we need to remember that considered moral judgments are not themselves immediately built into the natural world along with the evolutionary starting point of morality to be found in natural moral values. In this important regard, explicit moral oughts are not themselves deeply embedded in the natural biological world. Still, what EMR calls natural moral values are deeply embedded in the biological world, and they are tied to the responses of many different species of organisms through normative circuits of the kind Dewey meant to capture with his expanded concept of the reflex arc. In this sense, morality in the form of natural moral values goes into the natural world at one end to come out at the other end in the form of moral ought statements, assuming humans have evolved to be able to articulate, argue about, and agree on considered moral judgments in wide reflective equilibrium. Going back in the other direction, EMR also supposes that moral ought statements are tracking moral truth to the degree that they are tracking the natural moral values that were the natural inputs to the processes that eventually produced those very same ought statements. In this sense, EMR goes from is to ought and then back again, but at no point does EMR go directly from is to ought. Humanly explicit moral norms arise as part of a more general form of moral normativity, the moral normativity defined by natural moral values and the variety of moral capacities that develop in response to them. For explicit human moral norms, you need to add thought, language, reasoned argument, and social agreement. You also need the evolving cultural institutions that arise out of these processes, that is, basic social institutions that will allocate varying weights to varying moral considerations, and not necessarily in the same ways from one culture to the next.

In avoiding the is-ought fallacy, there is a related false dilemma that we also want to avoid. At its simplest, the dilemma arises as follows: does your form of evolutionary ethics tell us anything morally interesting? If so, it commits the is-ought fallacy. If not, it is morally irrelevant, and so we should just go back to doing philosophically informed ethics without worrying about the natural features of the world that make moral argument and agreement possible. According to EMR and our argument in this chapter, this is a false dilemma. It assumes too simple a relationship between empirical statements about natural moral values and ethical claims about what we ought to do, given the current state of our society's moral wide reflective equilibrium. The connection between the two kinds of statements is never going to be as tight as this dilemma takes it to be.

Tracking moral truth

On EMR's approach to morality, what goes into the natural world at the biological end is not what comes out at the other end of human efforts to reach points of moral wide reflective equilibrium. Natural moral values do not take us in a short and direct line to moral ought-statements, or more precisely, to considered moral judgments in wide reflective equilibrium. Learning more about natural moral values and their evolutionary development will not simply enable us to read off human moral truths from such biological results, as Wilson suggests we might be able to read off human moral truths from a deeper empirical study of the human limbic system. Nonetheless, natural moral values provide the underlying real aspects of the natural world that provide a crucial component of the truth conditions for the set of considered judgments in wide reflective equilibrium. To the degree that a set of considered moral judgments in reflective equilibrium tracks natural moral values, to that same degree can we suppose the judgments in this set to be true.

In his discussion of theory acceptance in ethics, Rawls (1971, 46–53) seems to suggest two ways of interpreting the idea of wide reflective equilibrium: constructivistically or realistically. These two interpretations of wide reflective equilibrium are discussed in Daniels (1980). On one possible realist interpretation of wide reflective equilibrium, an interpretation of wide reflective equilibrium suggested by Rawls' analogy between our sense of justice and our linguistic competence, what we are in effect doing in looking for the best overall ethical theory is tracking what our moral sense would tell us we ethically ought to do, were we to Socratically reflect on and potentially revise contentious items within our overall set of ethical beliefs. On a more constructivistic interpretation of wide reflective equilibrium, what we are in effect doing is deciding on what we should reasonably accept as being what we ethically ought to do, having agreed to a fair or otherwise appropriate procedure for making such decisions. In his third Dewey Lecture, Rawls (1980, 554–572) clearly opts for a constructivist interpretation of wide reflective equilibrium. What is important is the coherence of our considered moral judgments in wide reflective equilibrium, given the constraints of the decision-making process that we have chosen to jointly adopt. This coherence need not be supposed to be aimed at any sort of external reality that might make our moral judgments true or false, and it is not aimed at disclosing the underlying reality of the moral capacities that make it possible – the process takes us where it takes us, as we exercise these capacities in constructing reasonable rules to structure the basic institutions of our society.

The abductive nature of our argument in this book pushes us in the other direction, towards a more realist interpretation of wide reflective equilibrium. For the reasons we've explored in earlier chapters, we think that the real basis of morality is not simply to be found in our moral capacities themselves but in the natural moral values these capacities evolve in response to and along with, insofar as feedback loops may be established between evolving capacities and newly developing moral values.

EMR's more general account of moral oughts does not limit morality to humans or to human thought processes. Moral oughts are always linked to some kind of psychological capacity that is itself linked to moral values. Moral values are the kind of thing that they are, independent of how humans have evolved to respond to them. There may be better or worse ways to respond to moral values, and our species may not be as well positioned as it might be in this regard. If so, this limitation in our biological situation may or may not come to matter to us, and it may or may not be something we are able to correct. In any case, humans are not the ultimate measure of morality, even if we are at this point in time the only species we know of capable of articulating, arguing about, and agreeing to moral oughts under conditions that might (or might not) closely track the natural moral values produced through evolutionary processes.

As an innate capacity, our moral competence has its own internal biological limits. But it enables us to morally experiment with different social arrangements and to gain experiences about what furthers moral values and what diminishes moral values. This is at least part of what we are doing through processes of moral argumentation and social change. As we pursue the most coherent set of moral judgments we can obtain as we live and argue together, we learn more about the nature of both moral values and our human moral competence.

We are also tracking moral truth, more or less closely, depending on how successful our ongoing process of seeking considered judgments is. As with scientific theories, we do not encounter moral truths one at a time, individual proposition by individual proposition or individual moral judgment by individual moral judgment. The epistemologically important questions are how well our considered judgments cohere with one another, which of our predicted paths of social improvement prove to be actual improvements, and how we deal with anomalous moral judgments – judgments that seem right to us, but that are not a good fit with our current or socially evolving wide reflective equilibrium. If these judgments are mistaken, what is making them mistaken, and how might we best correct the degree to which they take us off what we might think of as our best course of moral thought and action? Can these anomalous judgments be accounted for as performance-based errors? Part of coming to grips with historical injustices is figuring out what enabled them to arise and what enabled them to be maintained. What did we think we were doing, and why? In his generally supportive discussion of Rawls' constructivistic interpretation of wide reflective equilibrium, Daniels argues against such a constraint on theory selection in ethics, but we think he is mistaken. Serious anomalies in science and ethics demand explanations. Other anomalies may fade away – the new theory might not solve or have to solve all the outstanding anomalies of the old theory. We may simply lose interest in some anomalies and their explanations. While we expect there are such explanations, it is no longer particularly interesting or important to pursue them.

At any given point in time, our considered judgments in wide reflective equilibrium will represent moral truth as well as we are able to know it. The extent to which any set of moral judgments tracks moral truth will ultimately be determined by the real moral values that processes of wide reflective equilibrium develop in response to. Moral truth as we understand it will be expressed in the set of judgments about what is morally best for us, given our current understanding of the moral values that matter to us in our current social and historical circumstances.

It is highly unlikely that there is a single, morally best way for our social institutions and relationships to be structured. But some structures may be morally better than others, and hence some considered judgments or sets of such judgments may be more on the track of moral truth than others. Like scientific truth, this is as good as moral truth gets for epistemically fallible creatures like us. Our theories, moral or scientific, may track the truths they are aimed at more or less closely, without ever finding any such truths in any absolute sense of the term.

To the degree that our moral obligations are rooted in our best understanding of what is morally true, where such truth depends not on what we merely believe to be true but instead on what is more likely to be on the track of independent moral truth, our obligations rest on something beyond us, something that we ignore at our peril. This something is the truth about the natural world we find ourselves in, whether we like it or not. As humans, we are free to ignore this truth. Sometimes it is ugly or otherwise unappealing. But experience suggests we will face it sooner or later, and that sooner is often better.

EMR tells us that there is a truth to morality that exists outside of our human efforts to come up with a coherent set of considered moral judgments in wide reflective equilibrium. If wide reflective equilibrium is on the track of this truth, and we act on the considered moral judgments supported by wide reflective equilibrium, we act as best as we can on the basis of moral truth. For humans, wide reflective equilibrium closes the loop between moral truth and moral normativity in a way that enables us to take this normativity as non-illusory. Although we can ignore moral oughts, this does not make the moral values they might be tracking disappear from the world we are part of. It may remain true that we ought to have paid attention to those values, perhaps precisely in the way that we were, prior to ignoring them. In this epistemological, metaphysical, and moral way, moral oughts may be unavoidable. The earth may yet move, regardless of whether we are willing to acknowledge it.

The authority of moral reasons

It is typically thought that moral reasons should trump other sorts of reasons, in particular reasons of self-interest. Are moral reasons in this way authoritative, according to EMR? One reason for supposing that moral reasons have a special authority over us, compared to reasons that come from subjective sources such as personal or social values, is that moral reasons seem to be grounded in objective truths that exist independently of whatever else we might think or believe to be true.

According to EMR, natural moral values do not immediately create moral reasons. Moral reasons emerge from processes of moral wide reflective equilibrium.

But some of the moral authority of moral reasons does come from natural moral values: like predators, natural moral values are extremely important things to pay attention to in the environments of highly cooperative and intelligent species. Ignoring them can be extremely dangerous to the individuals and groups in question. Even so, moral reasons are reasons for acting that emerge from moral argument and moral agreement, and as moral systems of thought become increasingly complex, moral wide reflective equilibria. Our commitment to these reasons as being decisive in how we ought to act comes most immediately from our joint commitment to participating together in these social and institutional practices.

Within this human context, some of the strongest threats to moral reasons will come from reasons of self-interest. But given the arguments of contractarians and contractualists, we might justifiably suppose that there will be a good deal of overlap between what is rationally good for each individual and what is morally good for all of us. For evolved creatures like us, it is unlikely that the dictates of self-interest and morality will be significantly opposed to one another when they are best understood within well-functioning processes of wide reflective equilibrium. On the other hand, it is also to be expected that there will always be tensions between what is good for the one (or his or her cohort) and what is good for the many. These tensions are built into the structure of our moral competence and will thus be an important part of what has to be worked out through ongoing processes of wide reflective equilibrium and different empirical experiences of moving too far in the direction of either partiality or impartiality.

For creatures like humans, whose lives matter greatly to them as the particular individuals they are, in special relationships with particular others, a central kind of moral question will be where to draw the lines between respecting oneself and the special others to whom one is closest and the more general forms of moral respect due to others to whom one is not so directly connected. As we argued in the last chapter, natural moral values will give no immediate answer to this difficult and pervasive sort of moral question. Where the lines are best drawn will depend on the ongoing arguments of the wide reflective equilibrium we are engaged in constructing for ourselves. What EMR adds here is the idea that while it is possible for the moral rules and social institutions in such an equilibrium to become too self-centred, it is also possible for them to become too group-centred. It might be argued that we could know this without appeal to natural moral values; to the extent to which this is true, EMR need have no quarrel with the point. But what EMR would emphasize is that to the degree that this point is true, it is on the track of deeper truths about morality more generally, not just human morality. If EMR is right, moral reasons ultimately rest on objective truths about the world that are directly related to the survival and flourishing of species that are social and intelligent. To the degree that this is so, we humans can reasonably take moral reasons to be authoritative.

But why should I be moral?

Self-interested and moral reasons may often overlap, and when they are in tension with one another, such tensions may be generally resolvable through the mechanism of considered moral judgments in wide reflective equilibrium with one another. What about cases where this is not so, as in the Ring of Gyges example?

On the view we are developing here, EMR does take moral oughts to be the more highly ranked kind of practical reason. Where to best draw the line between the oughts of rational self-interest and morality is determined through processes of wide reflective equilibrium that try to find the best balance between partial and impartial moral reasons in particular historical and social circumstances. Whatever the tensions between what is good for one (or some) and what is good for the many, the best way to resolve these tensions is through processes of wide reflective equilibrium guided by moral competence, reason, and our ongoing social commitments to one another. Such processes are also, according to EMR, on the track of moral truth, at least in those cases where they are moving us closer to the natural moral values that are at the foundation of both human morality and morality more generally.

But still: what if an individual capable of self-conscious reflection and free will wants to stand completely outside these processes and ignore their results? Is there any way to tell such an individual that he ought not to? In particular, is there a moral reason that he has to accept for telling him he ought not? The short answer has to be no: on our account, like many others in ethics, moral reasons come from within our considered judgments in wide reflective equilibrium. We should note that one can raise the sceptical question "but why should I accept reasons of this sort?" for rational self-interest and consistency as well. We might suppose it less likely to arise in such cases, but if it does arise, it is no more tractable than in the moral case. Given human free will, one can choose to ignore reasons of rational self-interest, consistency, or morality. Unless one has a pathological condition, probably not for very long, or to any very desirable effect, but one could still choose to act in completely arbitrary and capricious ways. In the moral case, one wouldn't be acting as one ought to, and along with Boehm, the rest of us might choose to ostracize or otherwise eliminate the sort of person we are considering here, although this is presumably not something that such a person would care about. Most of us, however, would care about this sort of thing and, indeed, we all ought to.

A scientifically positioned outside observer could also see that the sort of person we are considering here would not be behaving as he or she ought to. Biological patterns can always come apart in individual cases, which they sometimes do. If the patterns are important to survival and reproduction, such cases will always be limited in their appearance and persistence. If morally capricious individuals are a problem for morality, they are not much of a problem for morality on a view like ours. Biologically, moral normativity is connected to survival. Where the fit is particularly tight, we may expect the instincts to be particularly strong. Those who ignore natural moral values, like those who ignore predators, are unlikely to fare well in processes of natural selection. The biological world is fundamentally a hostile place for individual organisms.

Much more morally dangerous on any view of morality are socially powerful individuals or groups who do not care much for the interests of others, either most others or particular groups of others, such as Jews, blacks, or Palestinians. Our best response to such individuals is the best response currently on offer: to engage with them in social and political processes aimed at a wide reflective equilibrium that treats everyone with equal amounts of moral respect. In addition, EMR offers an important counterweight to claims that certain forms of social inequality are natural and inevitable. In addition to anthropology and history, we can add a deeper biological counter-argument against such claims of the naturalness and inevitability of human hegemonic social and political relations.

Even if our approach to morality cannot by itself help to move socially powerful individuals or societies in the morally appropriate directions that they ought to go in, we can at least offer important elements of an explanation for why this should be so. In addition to the sources of social or historical inertia that make it hard to undo oppressive social structures from the inside of those structures, our moral competence is individually developed in particular social circumstances, which may of course be oppressive in one way or another. What we feel to be natural balance points along the continuum that stretches between greater or lesser forms of partiality or impartiality will depend on the kind of society we are raised in, and in societies where they exist, in the kind of social class we are raised in. Given the social circumstances in which our individual moral competences develop, we may, too many of us, get stuck at a problematic balance point, weighting too heavily impartiality or more likely some form of partiality. The feedback loop between competence and moral judgments goes in both directions, feeding back into the formation of individual competence via the considered moral judgments we grow up with.

This process may lead us to moral dead ends: oppressive social circumstances from which our underlying competence cannot break us free. The successes of the feminist and abolitionist movements should give us some cause for optimism in this regard, even if the current levels of social support in some powerful countries like the United States for unregulated capitalism make us more worried about the possibility of neoliberal, individualistic dead ends that we may not be able to escape. Some forms of social power may simply become too powerful to undo from the inside, given the limits of our moral psychology and other social limits to the degree to which oppressive social structures can be challenged from within. Having eaten all our cake we may run out of bread, and so we may wind up with some advanced-capitalism equivalent of a bunch of lonely stone heads with their backs to the sea, at least the ones that do not get toppled or submerged by rising sea levels. A mixed bag of metaphors, to be sure, but one we hope has a clear point: when oppressive systems fall, so do heads. It may take a lot of fallen heads to threaten our species with extinction, but species flourishing is much more easily threatened than species survival.

Divine commands and moral twin earths

Having taken up the Ring of Gyges, we should also consider, against EMR, an evolutionary version of the Euthyphro dilemma. Just because natural moral values might ultimately be tied to the soundness of moral arguments that tell us, in a context of wide reflective equilibrium, that we ought to do something, does that really mean that we ought to do that thing?

In the *Euthyphro*, Socrates poses a dilemma for Euthyphro, who proclaims that moral goods are simply those things loved by the Gods. Socrates presses Euthyphro to say whether things are loved by the Gods because they are good, or good only because they are loved by the Gods. If the former, we have yet to say what makes good things good. If the latter, whatever the Gods may happen to love comes out as morally good, however bizarre the tastes of the Gods might turn out to be when it comes to what is morally good and what is not.

Because the Greek Gods were many and capricious, this was not an idle worry for Euthyphro. Yet if we were to say on behalf of a Christian God, for example, that he would certainly not love anything morally awful, like killing human babies for the fun of it, the problem remains. If what God loves is the sole and ultimate measure of morality, why could God not tell us exactly this, that it *is* morally good to kill human babies for the sheer fun of it? God is good, of course, but if all his goodness amounts to on this horn of the dilemma is his own love of self, there would seem to be no bounds to what he might tell us to do or not to do.

Our point here is not to rescue either the Christian God or the Greek Gods from the horns of the Euthyphro dilemma. The question for us is whether a version of the Euthyphro dilemma applies to EMR. If evolution had gone in different directions, could different moral values have resulted, moral values that might differ from or be exactly the opposite of the natural moral values that have been the focus of this book?

Our first point is that natural moral values do not by themselves tell us we ought to do anything: moral ought-claims come out of moral wide reflective equilibria. Even so, the truth of moral claims does ultimately come to rest on natural moral values. Could they have been other than they are? For EMR, this is a bit like asking whether lions could have been something other than lions. Kinds of things are the kinds of things that they are. If they were some other kind of thing, they would be some other kind of thing. Nonetheless, we can still ask: could something have evolved that was lion-like, without its being the kind of thing we currently believe lions to be?

In terms of species kinds, some other species could have developed from the node that created lions, either alongside lions or instead of lions. This makes it important to emphasize, again, that moral kinds, as EMR understands them, are not species kinds. They are deep structural features of environments containing organisms that are social and intelligent.

Given the sort of environments that natural moral values arise from, it is hard to see how they could be significantly different from what they are or at least appear to be at this stage of empirical investigation. For intelligent and social creatures, some things, like helping each other, would seem to naturally arise as the kind of thing that it is. And so too cheating, as the distinct kind of thing that it is. Both cheating and helping arise as possibilities in species after species, where these species are social and intelligent. Environments that leave open the possibility of large predators might or might not include lions, but moral values seem to be a more general and regular feature of biological environments than particular species like lions. Moral values seem like large predators in this regard: given the right sorts of environments, these general kinds of things will appear and have the general kinds of effects that they do.

The issue here might be understood as a more metaphysically subtle version of the moral twin earth argument of Horgan and Timmons (see Horgan and Timmons (1992, 2000) and Copp (2007)). The argument here is more subtle because of the distinction EMR draws between human morality and morality more generally.

In one version of the argument (Horgan and Timmons 1992), we are asked to imagine a human morality on earth where rightness is cashed out in consequentialist terms. On moral twin earth, twin earth humans also talk about moral rightness, but their usage of the term is cashed out by a deontological understanding of right and wrong. In terms of what humans on earth and twin earth humans call right and wrong, there is thus a good deal of overlap but occasional instances of apparent disagreement. In some cases, consequentialism will countenance lying where Kantian ethics will not, for instance. Is the moral disagreement real, or only apparent? If human rightness refers to actions that maximize the overall good and twin earth rightness refers to actions that respect all other rational creatures as moral equals, the disagreement is apparent: earth humans and twin earth humans are not ultimately talking about the same things. This seems counter-intuitive, the argument goes, and so we should conclude that moral terms like right and wrong do not refer to natural features of the world like maximizing the overall good or respecting the rationality of others.

From the point of view of EMR, the immediate problem with this argument is that human moral judgments, presumably drawn from two different wide reflective equilibria, are determining the ultimate references of moral terms. For EMR, biological theory, rather than moral theory, ultimately determines the references of moral terms, through natural kind terms that directly refer to things in the environment. Human morality is built on top of natural moral values through the cultural and argumentative evolution of considered moral judgments in wide reflective equilibria. If the equilibrium on earth is consequentialist while the equilibrium on twin earth is deontological, as long as both equilibria are stable, there is no real moral disagreement according to EMR if particular moral judgments vary across the two equilibria. Depending on the amount of variation, and perhaps the kind of variation, we might still wonder whether one (or both) of the equilibria is as stable as it might seem, especially once it is confronted by the other equilibrium.

The problem for EMR, however, is that a version of the twin earth argument might seem to recur at the level of natural moral values. Could the natural moral values identified by human biologists on earth differ from the natural moral values identified by biologists on twin earth?

According to EMR, natural moral values function in particular ways in regulating the behaviour of organisms that are social and intelligent. What is being selected for is a particular kind of functionality. Moral values are not just functional because they were selected for. What is emerging from selective processes are good kinds of things that play particular roles in the social interchanges of organisms for whom morality is possible, where morality is a particular way of regulating the behaviour of these organisms. What way is this? In general, a way in which the individual goods of the organisms are interrelated so as to minimize conflicts and maximize the integration of the separate interests of the separate organisms. It may be that for social and intelligent organisms, other natural values might play these roles – values other than the ones identified in earlier chapters.

Such a world might be biologically possible, but given the biological world as we know it, it is hard to see how. In our biological world, across species that are unrelated, the same sorts of natural moral goods seem to have the same general sort of function in regulating the behaviour of the members of these species towards one another. If there were a biological twin earth, this would be interesting and important in terms of our basic biological understanding of morality. But it would not mean that the terms of human morality did not ultimately refer to the natural moral kinds that were part of the biological development of life on the planet earth. It would mean that there was more to moral goodness, at a biological level, than we have so far been able to discover on our own planet, at our own current level of biological understanding. More empirically well-developed moral twin earth thought experiments might be useful for stimulating our biological imaginations, but they do not necessarily pose a deep philosophical threat to EMR as an account of human morality or as an account of morality more generally.

It is instructive to compare our account of the references of moral kind terms with the original twin earth example from Putnam (1973) in which "water" refers to H_2O on earth but to some other molecular structure XYZ on twin earth. Although it is hard to imagine what sort of an alternative molecular structure could give twin earth water exactly the same macro properties as water in a world that is otherwise exactly the same as ours, including the evolution of biological organisms that depend on water, it is logically possible that something like this might be so. Depending on the underlying science, if humans from earth got into a disagreement with humans from twin earth about what "water" really referred to, then it might be possible such a disagreement would only be a verbal one. Humans on earth survive on H_2O, humans on twin earth survive on XYZ, nothing more to be said. But the disagreement and the discoveries behind it might lead us not only to revise our idea of what water is but also (possibly and more deeply) a good deal of our thinking about chemistry and biochemistry. Lavoisier's discovery of oxygen, after all, sparked the chemical revolution. Depending on what is going on with XYZ and the evolution of life on twin earth, one might suppose a scientific revolution of similar magnitude might result from the scientific discovery of the twin earth as Putnam asks us to imagine it.

In any case, given the kind of pervasive, functional and normative role that moral values play in the lives of organisms that are moved by them, including those who are able to talk and argue, if humans from earth disagreed with humans from twin earth in a similar way over what moral goodness really referred to, the problem would likely have to be more than merely verbal. H_2O and XYZ may play the same functional roles in keeping humans on earth and humans on twin earth alive, but nothing follows about whether humans ought to drink XYZ or twin earth humans H_2O – presumably not, because neither would be hydrated by drinking the wrong fluid. But if humans from earth disagreed with humans from twin earth over whether helping others who needed it was a moral value, supposing everything else about them was exactly the same, this is something that would have to be worked out in terms of both sets of wide reflective equilibria. Moral values play a pervasive regulatory role in the lives of those who are subject to their normative force, and if two sets of humans are regulating their lives in the same sort of way, based on fundamentally different kinds of natural moral goods, this is something that would have to be worked out and accounted for.

As we are treating it here, the moral twin earth argument is a variant of the more general sceptical argument (e.g., Ruse (1986) and Joyce (2006)) that regardless of whether moral values really existed, we humans could be talking about morality in exactly the ways that we do. According to this sceptical argument, whatever is going on in moral discourse, reference to real-world moral values cannot be part of it. The general point of EMR is that semantic considerations do not determine whether or not real moral values exist. Whether or not they exist is an empirical question, not a semantic question. The mere logical possibility of a moral twin earth poses no immediate worries for EMR.

Searching for moral is-es among the moral oughts

Sceptical arguments are of more philosophical than empirical interest. EMR is at its core an empirical theory, and from the point of view of ethical knowledge and truth, it is one of many empirical theories that feed into the process of wide reflective equilibrium. It is, however, a particularly important empirical input into this process, because, linked together with the other empirical theories that are inputs to the process, it is an integrating part of what is making the ought statements that emerge as outputs from the process come as close to the truth as they do. Moral claims are ultimately claims about moral values, and while all moral values are hardly natural moral values, natural moral values are at the foundation of all claims with moral content. To the extent that such claims are tracking the moral truth, natural moral values are where moral truth comes ultimately to rest. They are the parts of our natural social environment that we either get closer to or further away from as we pursue a set of moral judgments in wide reflective equilibrium.

In this regard, there are two final points we want to make here. The first point goes back to the discussion of Rorty and Dennett at the end of our second chapter. We might worry that when it comes to ethics, there are no experiments to be performed. On the line of thought we are attributing to a pragmatic theory of truth as represented by Dennett's side of the argument, we know that our empirical theories are connecting with a world that exists independently of our theorizing about it because experiments turn out one way or another. This is what is ultimately telling us how closely (or not) our theories are tracking the truth about the empirical world. There are of course no instant successes or failures in this regard: experiments that turned out as we expected may in fact represent mistaken theories, and single experiments that failed may not tell us that the theory that generated them is mistaken. But over time, our theories are being measured by both coherence constraints and constraints of experimental testing.

In this regard, we side with Dewey's generally pragmatic approach to ethical knowledge and truth. In any society, we are engaged in ethical experiments on an ongoing basis. As part of ongoing processes of moral wide reflective equilibrium, we are arguing about how to best understand our current set of moral rules and institutional arrangements, as well as proposing and developing modifications, additions, or deletions to those rules and arrangements. In many Western democracies, we are now seeing, for example, an ongoing ethical experiment regarding the permissibility of euthanasia. Can the autonomy of competent patients be respected when it comes to requests for physician-assisted death at the end of life without endangering vulnerable individuals who might be killed when such a treatment option is not in fact in their best interest as patients? Time will tell, as we continue to develop and modify health care policies in this area. We now cease life-saving treatments for some patients who are not competent to make their own medical decisions, such as severely impaired newborns who are in pain and not expected to survive. Should we extend euthanasia to them, once we make it available to competent patients in what seem to be morally similar circumstances? Such an extension of existing policies allowing voluntary euthanasia would take us to policies allowing limited forms of non-voluntary euthanasia.

Our example here is complicated, and it is one among many. Our point is that moral experimentation is ongoing, and that it may take us closer to or further away from values like helping another when they need us to. Helping someone to die when they are suffering hopelessly and needlessly would seem to be an instance of this value, a value that starts as a simple natural moral value but becomes much more socially complicated as our societies become more socially complicated through the development of health care systems and professions like that of medicine. But regardless of however complicated our social institutions become, the suffering person either gets our help in ending his or her suffering or does not. Porpoises either arrive and buoy up a drowning swimmer or they do not. The ape either wades into the moat and saves the drowning toddler or it does not. Arguing and agreeing about what we ought to do matters to what we in fact do, as well as to whether what we do tracks the natural moral values that are at the foundation of human moral arguments and agreements.

Related to the question regarding experimentation is the concern that when we give up on an observational result in science because we have come to think it is mistaken, we often hold our new theories accountable in terms of being able to explain this mistake. Why did the observational result look right to us when it was in fact not? For example, if it turns out that the sun does not move across the sky when we view it from earth, why does it look to us like it does? We are prepared to accept the claim that our scientific theories are tracking the truth partly because our scientific observations are credible in ways that moral observations at any level of generality seem not to be. As Daniels (1979, 270) puts the point, although we regard our scientific observations as revisable, we take them to be reliably connected to the real world that they are observations of unless specific reasons arise to suggest that they are not. This is because we typically have theoretical accounts of why our observations are reliable, usually causal accounts linking our making the observations we do to the things that they are thought to be observations of. If we have to revise an observation statement, this means our account of why this was a credible observation in the first place was somehow mistaken.

Harman (1977, 6–7) makes a similar point about scientific observations. If you are observing a proton in a cloud chamber, our current theory about protons is explaining not just what you are observing, namely a proton, but that you are observing it, that is, that there is something in the cloud chamber, a proton, and that this proton is part of a causal explanation for your having made the observation that you just did. The proton hypothesis explains not only the content of your observation but also that you made the observation in the first place. If there are no protons, we then need to ask, what caused your apparent observation of one? On Harman's view of ethics, there is no good reason to suppose that we need to hypothesize real moral values to account for the fact that we make moral observations, like the one about the hooligans setting the cat on fire being morally wrong. All we need to suppose is that we have a certain sort of psychological make-up that leads us to make judgments about moral wrongness when we see things like hooligans setting cats on fire. We are set up to think that such things are bad, and so they are, but only because of us and our psychological capacities. These capacities are of course shaped by evolutionary development, but the kind of evolutionary development at issue for a view like Harman's is not one that ties the development of the capacities to particular features of the world that the capacities get more or less right. On a view like Harman's, it may pay, in terms of reproductive success, to pay attention to the screams of cats, but not because these screams are in and of themselves morally bad.

Against views like Harman's, EMR does directly tie the evolutionary development of moral capacities to the morally good and bad kinds of things that these capacities are tracking. So we need to ask, concerning the revisability constraint on observational statements in science, whether we hold our ethical theories to a similar standard. If not, why not? Our short answer here is that we do sometimes hold our ethical theories to exactly this constraint. If New World slavery turns out to be wrong, we want to know why we thought it was morally permissible for as long as we did. In other cases, we may be more lax. Somebody wins a moral argument against us and we revise our moral judgments. But the damage of the mistaken judgments was not serious, and the sources of moral error are as well known as they are numerous. We sometimes value our own self-interest more than we should or, similarly, the interests of those close to us. We are nepotistic, and we care about social standing, social hierarchies, and social power. We fail to consider as much as we might have how our actions are actually going to affect some other person. Moral errors are a regular occurrence in our lives, and often they are just that: errors that arise from a variety of common causes that are easily corrected.

Against such errors, what EMR offers is the possibility of truth-tracking causal explanations for at least some of the moral judgments that make up moral wide reflective equilibria. Natural moral values arise as real kinds of things in certain social environments. Evolutionary developments enable organisms in these environments to reliably detect these kinds of things. These detection mechanisms develop further into instincts, some simpler and some more cognitively sophisticated. These sorts of instincts, in humans, along with language and thought, enable us to create increasingly complicated sets of moral rules and increasingly complicated institutional arrangements with moral dimensions to them. This causal account of human moral judgments explains why we make such judgments and why these judgments have at least some of the morally important content that they do.

In science, observational and theoretical statements sort themselves out in more regimented ways than they might seem to in ethics, even assuming in both cases that all observational statements are theory-laden. It is harder to say, in the context of ethics, which moral judgments, exactly, are acting as observational statements. But Harman's example with the hooligans and the cat seems like a good place to start. Sometimes moral arguments come to an end with a telling example like this one, something that looks immediately wrong to us in a context that is not heavily theory-laden. Or to revert to another example, some of us are getting cucumbers while others are getting grapes for doing exactly the same task, with no other morally relevant features to distinguish our two groups. In terms of morally significant features of the world, the screams of cats or the tastiness of grapes compared to cucumbers are much less theoretically interesting than the structural elements of the situations into which they might figure.

In another of the many cat videos available online, one cat is in hot pursuit of another. The cats both run up a high tree and the fleeing cat plummets earthward. The attacking cat quickly descends from the tree and continues to attack the cat on the ground, although it has been clearly damaged by its fall and its responses are feeble. A group of monkeys, also in the video, has been closely watching the drama unfold, and at this point several of its members intervene to drive off the attacking cat – repeatedly. Perhaps their observations and actions are entirely devoid of moral content but perhaps not. It may be that unfair attacks are unfair attacks, and that they ought to be stopped if one is in a position to observe them and to stop them. The divide between fair and unfair attacks may be a real feature of the biological world, and it may be that this feature of the biological world is what is behind the selection of traits like those in monkeys and in us when we observe and respond to instances of the strong taking this particular kind of advantage over the weak. Or so we have tried to argue over the course of this book.

Bibliography

Copp, David. 2007. *Morality in a Natural World: Selected Essays in Metaethics*. Edited by Jonathan Lowe and Walter Sinnott-Armstrong, *Cambridge Studies in Philosophy*. Cambridge: Cambridge University Press.

Daniels, Norman. 1979. "Wide Reflective Equilibrium and Theory Acceptance in Ethics." *The Journal of Philosophy* 76 (5):256–282.

Daniels, Norman. 1980. "On Some Methods of Ethics and Linguistics." *Philosophical Studies: An International Journal for Philosophy in the Analytic Tradition* 37 (1):21–36.

Harman, Gilbert. 1977. *The Nature of Morality: An Introduction to Ethics*. New York: Oxford University Press.

Horgan, Terence, and Mark Timmons. 1992. "Troubles on Moral Twin Earth: Moral Queerness Revived." *Synthese* 92:221–260.

Horgan, Terence, and Mark Timmons. 2000. "Copping Out on Moral Twin-Earth." *Synthese* 124:139–152.

Joyce, Richard. 2006. *The Evolution of Morality*. Edited by Kim Sterelny and Robert A. Wilson, *Life and Mind: Philosophical Issues in Biology and Psychology*. Cambridge, MA: The MIT Press.

Mackie, John Leslie. 1977. *Ethics: Inventing Right and Wrong*. Harmondsworth: Penguin.

Putnam, Hilary. 1973. "Meaning and Reference." *The Journal of Philosophy* 70 (19): 699–711.

Rawls, John. 1971. *A Theory of Justice*. Cambridge, MA: Harvard University Press.

Rawls, John. 1980. "Kantian Constructivism in Moral Theory." *The Journal of Philosophy* 77 (9):515–572.

Ruse, Michael. 1986. *Taking Darwin Seriously: A Naturalistic Approach to Philosophy*. Oxford: Blackwell.

Wilson, E.O. 2004. *On Human Nature*. Cambridge: Harvard University Press. Original edition, 1978.

Conclusion

The issues raised at the end of the last chapter are significant for EMR as an empirical approach to morality. Our short discussion of these issues is not meant to be conclusive. Better responses to such issues will require more empirically and morally well-developed versions of EMR. Our purpose in this book has been more limited in scope. Our argument has been that EMR offers a plausible and potentially powerful explanation of the natural origins of species-independent and biologically real moral values. EMR merits further development as an empirically based theory of morality.

Part of what we have tried to do in this book is to level the playing field against alternative empirical and philosophical accounts of the phenomena in question. Humans appear to make moral judgments, and the question is how best to explain this feature of the empirical world. Most other approaches to morality suppose that whatever it is that is going on with these judgments, they are not tracking natural features of the biological world in a way that could be directly connected to moral truth. EMR hypothesizes that there is such a connection, and we think there are interesting reasons to suppose that this general hypothesis might be on the right empirical track. Of course, we still need to show that EMR provides the best explanation of the biological phenomena: but so too do other theories of morality, at least those theories with some degree of empirical content. Claims that moral values do not really exist, or that they exist as non-natural but nonetheless real things accessible only to human cognitive capacities, represent alternative hypotheses that also need to be empirically worked out much more carefully than they currently are. We cannot simply suppose that they are more likely to be true because a naturalistic approach to moral realism seems to assume far more than it needs to.

In this regard, there are two main lines of objection to EMR that we have addressed in the course of the book. The first objection is that there seems to be nothing in the real world to which moral terms might directly refer. The first four chapters of the book are our response to this objection. It is not implausible to suppose that there are regularly recurring features of certain kinds of environments that are driving the development of moral or at least proto-moral capacities. These capacities have the general structures that they do at least in part because the environments they evolve in possess the general structures that they do. The second objection is that there is no reason to suppose that if they exist, these general environmental structures might be morally significant. The last four chapters of the book are our response to this second objection. The natural kinds we identify in the first half of the book seem to be at the core of what we humans are arguing about, when we argue about morality.

Considered by themselves, each of these two lines of response might seem dubious. If the first half of the book is right and there are naturally good things of the general kind that EMR hypothesizes, why suppose these goods are moral goods? If the second half of the book is right, certain features of our social environments may cause us to change our moral beliefs, but why suppose that this is in response to something morally significant about such features of the world in and of themselves? It is when we put the two halves of the book together that a plausible case for EMR as an empirical theory of morality emerges. The specific features of the human social environment that are tied to moral responses in humans are the same general kinds of things other species respond to in similar kinds of ways. Biologically, morality seems to start simple and to get more complex. Human morality is special but as a special case of a more general biological phenomenon.

We have said throughout the book that EMR is an empirical theory. If it merits further development, in what directions might we look for additional evidence and for more specific hypotheses? Neither of us is an experimental scientist, so an important part of what we take ourselves to be doing in this book is opening up a promising avenue of theoretical speculation to the imaginations of those individuals who are experimental scientists. Even so, we can suggest some avenues of further development that seem to us to be more immediately interesting.

First, there is the question of how well the central hypothesis of EMR might be deployed in integrating the evidence that is currently available to it. What might a general catalogue of environmental features look like, of the kind or kinds that EMR calls natural moral values? Across species that are social and intelligent, what are the differing sorts of opportunities for helping others in trouble, responding to others in distress, and treating others fairly in exchanges of goods or favours? To what degree do these features of their particular worlds figure into the Umwelts of each species in question? Which of these features do these Umwelts pick up on, and to what degree, and to which of these features do they seem oblivious? Do the morally significant features of the environments in question, along with other relevant features of the environments, genuinely help to explain the evolutionary development of pro-social capacities of the species in question? EMR's central claim is that there are general environmental features that are realized differently across different environments and that these similarities and differences account for the similarities and differences in the pro-social psychological traits that we observe across a wide diversity of species. To what degree does this hypothesis play out, as we more carefully sift through the evidence currently available to us?

In terms of new evidence, one important sort of thing to look for might be pro-social analogues of the beaks of finches. Are there closely related species that started as a single species, subgroups of which got separated and then got different, where morally significant differences in traits can be directly connected to morally significant differences in the separate environments? Did getting separate, in other words, lead to environments with differences in opportunities to help, empathetically care, and to interact fairly with others? And if so, do these differences in the separate environments help to explain the differences in the relevant traits among the species in question?

If we are indeed talking about the natural kinds at the centre of EMR as moral kinds, we also need to look more carefully for human evidence that bears upon their existence and causal efficacy. Are there more cases like the slavery one, where a close reading of the historical record uncovers what EMR calls natural moral values, as important but unnoticed or underappreciated causal elements in the shift in moral thinking that distinguishes such cases as being of historical and social significance? If so, how well can the presence of such values be connected to the forms of human moral reasoning also likely to be involved in these progressive episodes of historical change? Are they well connected to the changes in the social contracts that seemed justified to the social actors who were arguing about the legitimacy of these changes, relative to the status quo? If part of what was going on was social argumentation in search of a more consistent set of moral judgments in wide reflective equilibrium, how well connected were the natural moral values in question to the greater consistency of the equilibrium in question? Here it is important to remember that the natural moral values hypothesized by EMR are not meant to supplant human forms of moral reasoning but to augment them. The claim in Chapter 5 was not that social contract thinking or human reason in general has no role in the search for moral truth but that by themselves they do not offer us a complete account of either moral justification or moral truth. For that we need natural moral values, and so the empirical question at the level of progressive social change is always over the degree to which such values may have been lurking, relatively unnoticed, in the background of such changes.

Finally, with regard to Chapter 7, it will be interesting, to say the least, to see how our current global situation plays out. Building walls to isolate groups of people from one another seems to be a prime example of the kind of performance-based error that was at the centre of our chapter on the human capacity for morality and what we called our human moral competence. If EMR is right, our moral competence is pliable enough to continue to enlarge the groups of individuals to whom we think we ought to offer help, empathetic caring, and fair treatment. Assuming we survive the current threats to our survival represented by continued nuclear proliferation, economic inequality, social strife, global warming, and generalized ecological collapse, it will be interesting to see if later humans come to regard our current fixation on building walls as a performance-based error. While such walls may have looked justified to us at the time as reasonable forms of moral partiality, this form of partiality, in the circumstances in question, was not truly reasonable.

As a final objection to EMR, we might consider the point that we could come to realize our moral error here without the help of EMR. But what makes us think this? EMR is not meant to provide the whole reason we might come to revise deeply held moral views as being mistaken. It is meant instead to connect our getting things morality right with the fact that other organisms may be getting things morally right as well, at least enough of the time for them to survive and flourish. That we humans move in the moral directions that we do is not biologically disconnected from the moral movements of other species that are social and intelligent, however intelligent they might be. EMR may be right, whether we come to realize this or not. We may be special, but maybe not that special.

Index

abolition 109, 125–127; *see also* slavery
adaptive links 7, 12
affective morality 92
aggression 133; and play 28–29, 46–47; and reconciliation 47–48
alloparenting 68–71, 90
altruistic punishment 49
anger 32, 62, 131
apes 68–69; *see also* chimpanzees; monkeys
appeasement 47; *see also* reconciliation
attachment 52–53, 55–56; *see also* maternal care; parenting
autonomy 45, 88–89

bees 24–25, 28
behaviours 102–105
Brown, John 122–124
Butler, Benjamin 118

capitalism 110, 112
caring 120–125; *see also* alloparenting; attachment; empathy
cheating 73; *see also* fairness
chimpanzees 62–63; and aggression 47–48; and the common good 65–66; and empathy 54; and punishment 50; and reconciliation 133; *see also* monkeys
Civil War 118–120
Clarkson, Thomas 113–114
colonies 23–24, 25
conscience 88–92
contractualism 96–102, 125–126, 156; *see also* social contracts
cooperation 3, 17, 19–20, 31, 92–94
cooperative breeding 68–69

Darwin, Charles 25
Dawkins, Richard 99–100, 152
deceit 22–24; *see also* fairness
de Waal, Frans: and conciliation 72; and empathy 49–51, 54, 77–78; and fairness 21; and moral instincts 61–63; and reconciliation 47–48
Dewey, John 33, 39, 152
dialogue 131
disgust 32
dodging 74–75
dogs: and fairness 21–22; and reconciliation 48
Douglass, Frederick 115, 120–122, 123

Emancipation Proclamation 119
Emerson, Ralph Waldo 125
empathy 29, 31, 50–55, 125–126, 136–137; *see also* caring; pain
empiricism 167–168
evolutionary trajectories 20–21, 24–26
experimentation 163

fairness 18, 21–22, 44, 62, 68, 90; and play 28, 46–47; and reason 96; *see also* cheating; deceit; justice
forgiveness 47, 48–49; *see also* reconciliation
fundamental attribution error 132, 138

Garrison, William Lloyd 115, 117, 122
gender 95, 133, 135–136, 139
genes 99–102
genetics 8–9
Glaucon 89, 91, 94
globalization 140
guilt 91; *see also* shame

Harman, Gilbert 6
helping others 42–44, 112–120
Hobbes, Thomas 96

impartiality 55–57, 133–141
individualism 79–81

individualism 93
in-group loyalty 11, 15, 32, 140–142

justice 31, 73, 96–97, 153; *see also* fairness; wild justice

kin selection 8, 24–25, 101–102; *see also* in-group loyalty

language 10, 13; and moral goods 39–40
light detection 26
Lincoln, Abraham 117, 123–124

maternal care 68–71; *see also* attachment
mistakes 19
monkeys 133; and conscience 90; and fairness 21–22, 44; and moral instincts 61–62; and pain 26–28; and parenting 69; *see also* apes; chimpanzees
Moore, G. E. 39
moral authority 88–96
moral capacities 9, 26–29, 31–32, 75–76, 105
moral codes 2
moral competence 129–130, 154, 168–169; and in-group loyalty 140–142; and moral performance 132–140; and naturalistic fallacy 144–145
moral emotions 6–7, 10; *see also* caring; empathy; fairness
moral environments 14
moral goodness 19–22
moral goods 1–8; examples of 42–44; and genetics 8–9; and the interests of others 40–42; origin of 9–12; as real things 39–40
moral instincts 9, 30, 61–68, 102–105
moral intuitions 40
morality 4–5, 71–72
moral kinds 105–106
moral normativity 33–36, 44–50, 147–150
moral oughts 9–10, 11, 149, 162–165
moral perceptions 130–132
moral performance 132–140
moral progress 77–81; and abolition 125–127; and caring for others 120–125; and helping others 112–120; and moral values 109–111
moral reasons 155–156
moral reflective equilibria 10, 12
moral sense theories: and maternal care 68–71; and moral instincts 61–68; and moral progress 77–81; and moral truth 60; and wild justice 71–77
moral sentiments 77–81
moral trajectories 30; and empathy 50–55; and impartiality 55–57; and moral goods 39–44; and moral normativity 44–50
moral truth 35, 60, 84–88, 153–155
moral values 1–2, 40–42; and evolution 17–19; and evolutionary trajectories 24–26; and moral capacities 31–32; and moral normativity 33–36, 44–50; and moral progress 109–111; as natural kinds 29–31

naturalistic fallacy 144–147, 150, 153–160
nomological danglers 85–86
numbers 84–85
nutritional goods 20, 21–22, 24–25, 34

observations 163–165
oppression 72–73
oxytocin 53

pain 26–28; *see also* empathy
parenting 68–71, 135
patrols 65–66
photographs 120–121
Piaget, Jean 102–103
play 3, 28, 46–47
pragmatism 35
predator-prey relationships 18
predators 2, 31
punishment 20, 73, 101–102; *see also* altruistic punishment

race 95, 109–110, 117; *see also* slavery
rats: and empathy 29, 54–55; and robbing 74–75
reason 13, 56, 80, 83–84; and contractualism 96–99; and moral authority 88–91; and moral truth 84–88; and self-interest 91–96
reasonable refusal 97–99
reciprocal altruism 8
reconciliation 47–48
reflex arc 34
religious beliefs 19, 116–117, 159
responsibility 45
Ring of Gyges 157, 159
robbing 74–75
rules 78–79

selection 6
self-interest 91–96, 110; *see also* capitalism

selfish genes 99–102
shame 88, 90–91; *see also* guilt
shrimp 22–24
slavery 109–111, 125–127; and caring 120–125; and helping others 112–120
social contracts 67, 94–95, 125–127; *see also* contractualism
survival 18
sympathy 90; *see also* caring; empathy

technology 35
termites 25
Thrasymachus 91, 94
trait selection 6, 8, 9, 14, 29, 101
tribal loyalty *see* in-group loyalty
trust 1, 19–20, 29; and empathy 50–55; and parenting 70
tuning 101, 104

Umwelts 63–64

view from nowhere 26–27, 85–86

war 10–11, 15
wide reflective equilibria 80, 87–88, 90, 93, 113, 130–132, 141–142, 147–149
Wilberforce, William 115
wild justice 71–77; *see also* justice
women 131; *see also* gender
Word, the 5
World War I 10–11